Lessons Not Learned

Lessons Not Learned

THE U.S. NAVY'S
STATUS QUO CULTURE

Roger Thompson

NAVAL INSTITUTE PRESS
Annapolis, Maryland

Naval Institute Press
291 Wood Road
Annapolis, MD 21402

© 2007 by Roger Thompson

All rights reserved. No part of this book may be reproduced or utilized in any form or by any means, electronic or mechanical, including photocopying and recording, or by any information storage and retrieval system, without permission in writing from the publisher.

Library of Congress Cataloging-in-Publication Data
Thompson, Roger, M.A.
    Lessons not learned : the U.S. Navy's status quo culture / Roger Thompson.
      p. cm.
    Includes bibliographical references and index.
    ISBN-13: 978-1-59114-865-4 (alk. paper)
    ISBN-10: 1-59114-865-0 (alk. paper)
    1. United States. Navy—Operational readiness. 2. United States. Navy—Personnel management. I. Title.
    VA58.4.T495   2007
    359'.03—dc22

                                                              2006100100

Printed in the United States of America on acid-free paper ∞

14 13 12 11 10 09 08 07   9 8 7 6 5 4 3 2
First printing

Dedication

THANKFULLY, AND very reassuringly, there are a great many U.S. Navy officers (serving or retired) who are willing to speak about their navy's failings. I submit that these men and women are the true patriots, not the credulous and defensive "Everything's just fine, we're the best, thank you" types who populate the Brobdingnagian U.S. military-industrial complex, the Pentagon spin-doctors pumping out warmed-over double-talk, and all others who cannot see the reasonable forest for the trees. "'The Navy as always,' Truman believed, 'is the greatest of propaganda machines'" (Michael T. Isenberg, *Shield of the Republic: The United States Navy in an Era of Cold War and Violent Peace*, vol. 1 [New York: St. Martin's, 1993], 61).

These reformers and thinkers try to make a difference, and they are the ones who are truly loyal, for they realize that one does not need to be an unquestioning reactionary to be a loyal and effective officer or sailor. One will find such men and women in the pages of the U.S. Naval Institute *Proceedings* from time to time, but the most influential in these ranks are such men as the late Adm. Elmo Zumwalt, Adm. Stansfield Turner, the late Rear Adm. Eugene Carroll, Capt. Dean Knuth, the late Scott Shuger, and former F-14 radar intercept officer Jerry Burns, all of whom are quoted herein. To these men, and the men and women like them now in the U.S. Navy, I respectfully dedicate this endeavor. You have heard all the hype about the U.S. Navy, I am sure, so this study will give you the other side, the side that does not often make it into the mainstream media, or the U.S. high school textbooks.

"As far as his comments in general, he feels that the Navy systems are oversold, overpriced, and undercapable. He is generally more pleased with the Air Force, but sprinkled criticism of us rather freely" (Maj. Gen. Perry M. Smith, USAF [Ret.], reading his notes on a 1974 job interview with Secretary of Defense Dr. James Schlesinger. From Perry M. Smith, *Assignment Pentagon: How to Excel in Bureaucracy* [Dulles, Va.: Brassey's, 2002], 176).

Contents

	Foreword *by Dr. P. Andrew Karam*	ix
	Acknowledgments	xiii
	Acronyms	xv
1	Introduction and Objective (*Quaere Verum*)	1
2	The "Exercises Aren't Real" Argument: My Riposte	8
3	David *vs.* Goliath: Diesel Subs and Mines Take On the U.S. Navy	15
4	ASW: A Low Priority?	40
5	A Lucky Break at Midway and the Big-Carrier Navy	63
6	The Russians Mug the *Kitty Hawk*, the *Saratoga*, the *Constellation*, the *Carl Vinson*, and Others	81
7	The Chinese: Know Thy Potential Enemy	94
8	Lax Security	97
9	A Few Realistic Men	101
10	This Isn't *Top Gun*—and Watch Out for the Little Guy	108
11	Lack of Training, Overrated Technology, Bad Policies, and Technocratic Leadership	138

12	Morale Issues, Racism, Drugs, Sabotage, and Related Matters	158
13	What Tom Clancy Does Not Know or Won't Tell You	168
14	Misleading Congress, and a Cultural Explanation	176
15	Conclusion	179
	Afterword *by Col. Douglas Macgregor*	181
	Appendix: USN Ships That Have Been Theoretically Destroyed	185
	Notes	187
	Bibliography	217
	Index	245

Foreword

by Dr. P. Andrew Karam

DECEMBER 7 IS A DAY FREIGHTED with emotion and memory. Although I am too young to have direct memories of the attack on Pearl Harbor, I visited the Arizona Memorial when my submarine pulled into Pearl Harbor in 1989. I was surprised to feel tearful at the thought of the bombs, the explosions, and the men dying as they struggled to respond. That Pearl Harbor was a tragedy is undeniable. That it was avoidable is also undeniable.

On December 7, 2001, I had another emotional experience—graduating with my doctorate after fourteen years of working and studying full-time, following my discharge from the U.S. Navy (I had spent eight years in the nuclear power program and five in the reserves, leaving as a chief petty officer). I feel very strongly that I owe a large part of my success to the discipline I learned in the Navy, and I know that a major reason I was able to stay up late for the years needed to complete my research and dissertation is that my memories of the long hours of work on the submarine reminded me that I was capable of this level of effort. I am certain that, were it not for my experience in the Navy, I would not have my current degree or my current career.

My experience is not unique—many ex-Navy "nukes" have similar stories. And that is part of the point of this reminiscence. The Navy is full of people with energy, intelligence, and drive. Many of them join the Navy to expand their horizons, to learn, to obtain job skills, to see the world, and more. From what I saw, many of the people in the Navy can and will do anything if they are given the time, resources, and leadership. No matter how good the Navy's ships, missiles, radars, and torpedoes, without good sailors who are properly trained and led, these weapons are worthless.

Although I did not particularly enjoy the time I spent in the Navy, I was proud of what we accomplished. Between 1986 and 1989, I made four "special operations" off the coast of the Soviet Union, and my submarine, the USS *Plunger* (SSN 595), returned from each one with valuable information about the Soviets. During my time on board, *Plunger* earned a Meritorious Unit Commendation and a Navy Unit Commendation—not bad for an old boat that went into the shipyard for decommissioning immediately after returning from its final op.

In spite of the good work we did, and in spite of *Plunger*'s successful record against the Soviets, I had many reservations about the Navy. These went beyond bad food, insufficient sleep, and some bad officers. It concerned me, for example, that submarine skippers received more training on the nuclear reactor plant than they did on strategy and tactics. It concerned me that all of our drills and training were aimed at passing inspections rather than stretching our understanding of submarining. It concerned me that when helping our surface fleet practice antisubmarine warfare, we were often instructed to help them to find us rather than challenging them to improve their skills. And it concerned me that the system for assigning personnel was so inflexible that the Navy would see competent, highly trained sailors leave rather than let them explore other technical specialties.

When I was in the Navy we faced an enemy who had an enormous edge in numbers, and our only hope of survival and success lay in superior technology and superior tactics. My experience was that we had these advantages. I was fairly certain that if we'd been involved in a shooting war with the Soviets, we would have won. But I suspected that the cost would have been higher than it needed to be, because I was not sure that the enemy would be as inept as we expected them to be. After reading this work, I am even more certain that while we probably would have prevailed in a conflict against the Soviets, it would have come more dearly than we would have liked.

It is possible to be respectfully critical of something (or someone) that you admire, in spite of its flaws. As a professor, I often find myself goading students who are talented but either don't recognize their potential or who choose not to exercise it. Challenging them, prodding them, sometimes even embarrassing them a little can work wonders; all professors have stories of students who, when all was said and done, were surprised how well they could do. That is the value I see in this work. Roger Thompson shows both a deep regard for the U.S. Navy and an appreciation for its potential.

At the same time, he has come to understand many of the factors that have kept, and continue to keep, the Navy from realizing this potential. It is easy to read this work as a negative polemic against the U.S. Navy, and I am sure that many will regard it in this manner. I disagree. If Roger had only criticism of the U.S. Navy, there would be no need for this work; it would suffice to simply let the Navy disintegrate on its own. But, as a professor, I know there is little as frustrating as a student who has the potential to be world-class but simply lacks the motivation to do so. Such students are the most challenging and the most frustrating—and ultimately the most rewarding. Roger has put an enormous amount of time, effort, and energy into the thoughtful work you now hold; let us hope it will be rewarded by, someday, seeing the Navy working at its full potential.

P. Andrew Karam, Ph.D., CHP
Assistant Professor,
Rochester Institute of Technology

Acknowledgments

I WOULD LIKE TO THANK Dr. Andy Karam, former U.S. Navy nuclear submariner and author of the book *Rig Ship for Ultra Quiet;* Capt. John L. Byron, USN (Ret.), former nuclear submarine commander; Dr. Robert Williscroft, former U.S. Navy nuclear submarine officer; Commander Peter Kavanagh, Canadian Forces (Ret.), a diesel submarine skipper, for providing details on his successes against U.S. nuclear submarines; Col. Douglas Macgregor, USA (Ret.), author of the book *Breaking the Phalanx;* Maj. Donald Vandergriff, USA (Ret.), author of *The Path to Victory: America's Army and the Revolution in Human Affairs;* Lt. Col. David Evans, USMC (Ret.), former military correspondent for the *Chicago Tribune*; Rear Admiral Fred Crickard, Royal Canadian Navy (Ret.); Jon E. Dougherty, investigative journalist and former U.S. Naval Reserve sailor; Squadron Leader J. R. Sampson, Royal Australian Air Force (Ret.); Henrik Fyrst Kristensen, for his critiques and for translating Swedish materials; Steve Cook; Mike Sparks; Carlton Meyer, former USMC officer and editor of *G2mil* magazine; Brigadier General (Dr.) Jaime Garcia Covarrubias, Chilean army (Ret.), for Spanish translation; Professor (Lieutenant Colonel, Reserve) Dmitry Pozhidaev, Soviet/Russian army (Ret.), for research assistance and Russian translation; and Dr. Emilio Meneses, for much information on exercises between the Chilean air force/navy and the U.S. Navy. I am indebted to all these fine people for their input, comments, suggestions, and constructive criticisms of earlier versions of this book. I would also like to thank Capt. Dean Knuth, USNR (Ret.) for background information on the sinking of two aircraft carriers in Exercise Ocean Venture 81 and for reviewing the section titled "David vs. Goliath"; and Col. Everest Riccioni, USAF (Ret.), the father of the F-16 fighter program, Lieutenant Colonel Pierre Rochefort, Canadian Forces (Ret.), and Lieutenant Colonel David Bashow, Canadian Forces, author of *Knights of the Air: Canadian Fighter Pilots in the First World*

War, for their advice on fighter combat; Major Lew Ferris, Canadian Forces (Ret.), and Major Leif Wadelius, Canadian Forces (Ret.), for their advice on ASW matters; Lieutenant Commander Aidan Talbott, Royal Navy, for his comparisons of the U.S. Navy and the Royal Navy; and Captain Jan Nordenman, Royal Swedish Navy (Ret.), for information on Swedish diesel submarines. My special thanks also go to Dr. Debora Shuger of the UCLA English Department, who kindly gave permission to use her late husband Scott Shuger's unpublished book manuscript *Navy Yes, Navy No*. Finally, I offer my thanks to all my other sources, who will remain safely anonymous, for their generous assistance, and to my late father, Major Reg Thompson, Royal Canadian Air Force/Canadian Forces (Ret.), for providing the inspiration to speak out against injustice and military stupidity and to "be my own man." However, the opinions stated herein are mine and mine alone.

Acronyms

AAA	antiaircraft artillery
ACM	air combat maneuvering
AFQT	Armed Forces Qualification Test
AMRAAM	Advanced Medium-Range Air-to-Air Missile
ANG	Air National Guard
ASUW	antisurface warfare
ASW	antisubmarine warfare
ASWTNS	Antisubmarine Warfare Tactical Navigation System
BVR	beyond visual range
CAP	combat air patrol
CETS	Contractual Engineering Technical Service
CF	Canadian Forces
CIC	Combat Information Center
CSG	carrier strike group
CO	commanding officer
CVN	nuclear-powered aircraft carrier
DDG	guided-missile destroyer
DOD	Department of Defense
FASWC	Fleet ASW Command
IJN	Imperial Japanese Navy
INSURV	Board of Inspection and Survey
NAS	Naval Air Station
NATO	North Atlantic Treaty Organization
OECD	Organization for Economic Co-operation and Development
OPFOR	opposing force [simulated]
OPSEC	Operational Security
RAAF	Royal Australian Air Force

RCAF	Royal Canadian Air Force [now Canadian Forces Air Command]
RCN	Royal Canadian Navy [now Canadian Forces Maritime Command]
RDF	radio direction finding
RN	Royal Navy
SAM	surface-to-air missile
SAT	Scholastic Aptitude Test
SOSUS	Sound Surveillance System
SSBN	nuclear-powered ballistic missile submarine
SSK	diesel-electric submarine
SSM	surface-to-surface missile
SSN	nuclear-powered submarine
UDT	Underwater Demolition Team
USAF	U.S. Air Force
USMC	U.S. Marine Corps
VID	visual identification

1

Introduction and Objective (*Quaere Verum*)

> I never did give them hell. I just told the truth, and they thought it was hell.
>
> —Harry S. Truman[1]

LET ME BEGIN BY stating that the U.S. Navy is an important fighting organization, but it is not a person. It is not the flag, and it is nobody's mother or child. It is an employer of hundreds of thousands of people, but importantly, it is one that has extracted billions of dollars from the taxpayers. It is not a religion, it is not sacred, and as such, it can and must be subjected to rigorous criticism when warranted. It is in the spirit of sincere and constructive criticism that I write this. I say this because, despite good intentions and extensive documented evidence, often provided by current or former U.S. Navy officers who want to turn this organization around, there are some who are apparently incapable of engaging in constructive but intellectually honest discussion on their current or former service. To these folks, the U.S. Navy *is* America, and to criticize the former is to mock the latter. I dismiss this paradigm, along with any and all counterarguments that are based on emotion, hyperbole, willful ignorance, or fideism; that invoke ad hominem abusive, the ad hominem circumstantial, *ignoratio elenchi* tactics; that are nonspecific and undocumentable (in other words, based on assumed facts that are not in evidence, better known as the old "I think you took these statements out of context, but I cannot rebut them because I do not know the actual context, and basically I do not like your argument so I am just grasping at straws to deflate it" gambit); and those based on disingenuous and unauthenticated contumacy or prevaricating bromides that do not wash in terms of reality, common sense, or precedent.

In this age of rampant jingoism in the United States, in which even the most thoughtful and well-reasoned criticism of the U.S. military is sometimes inexplicably equated with contempt or polemical disrespect, some reactionaries might even go so far as to claim a work such as this must ipso facto be "anti-American." Indeed, Michael Parenti said recently, "With the link between militarism and patriotism so firmly fixed [in America], any criticism of the military runs the risk of being condemned as unpatriotic."[2] Come to think of it, some "buffs" out there, some of whom are not even Americans or have not served in the U.S. Navy and therefore should have no emotional attachment to this organization, might react badly to this argument too. For them, I would suggest entering this book with an open mind, allowing verifiable facts the space they deserve, and I would ask, if any of what I share below can be shown to be inaccurate, that they contact me with corrections. In any case, I eschew this simplistic, linear thinking that criticizing the U.S. Navy is America bashing; to those who do not, I would point out that seven of my relatives have served in the U.S. Navy. Two of them served on the destroyer USS *Arnold J. Isbell,* one on the escort carrier USS *Rudyerd Bay,* and another was a parachute rigger with Attack Squadron 65 on the supercarrier USS *Dwight D. Eisenhower.* Other relatives served in the U.S. Army, Marine Corps, and Air Force in World War II, Korea, and Vietnam. I am pleased to mention that one of my relatives was born at U.S. Naval Hospital San Diego. Quite obviously, I have no desire to offend my American relatives. On top of all that, my late father served for a time as a foreign exchange pilot with a U.S. Navy patrol squadron, VP-31 ("The Genies"), out of Moffett Field Naval Air Station, where he qualified to fly the Lockheed P-3C Orion.

I also offer much praise for other branches of the U.S. military, especially the Air Force, for their relatively high level of professionalism, selection standards, and excellent aircraft, especially the F-16. Note well, however, that I do not place the Air Force on a pedestal, and to prove it, let me say from the get-go that while it is better than the U.S. Navy in some ways, it too suffers from many of the same afflictions, including a bloated, overspecialized enlisted force, reliance on expensive, poorly tested weapons, and an "up or out" promotion system, to name a few.

To borrow a phrase from a well-known Jack Nicholson movie, if "you can't handle the truth or are one of the many who are "blinded by hype about our technological and ethical superiority,"[3] then I suggest, respectfully, kindly, and sincerely, that you go no farther. No one should take what I am about to say personally. Besides, if you disagree with my thesis, and if the U.S. Navy's way of doing things is somehow validated in a future war, and without too much "dumb luck," then you have nothing to worry about—and hence, nothing to be angry about, either. If I am right, however, you have reason to be angry—at the U.S. Navy, the Pentagon, the Congress, the president, and defense contractors—but not me, for I am merely the narrator, and I will be kind enough not to say "I told you so." As I always say, it is the mark of a true scholar and gentleman (I used the term "gentleman" without regret, as there is no comparable "gender-neutral" expression) to be able to disagree respectfully and courteously. A true gentleman scholar is also not afraid of mere ideas, understands the precepts of civilized discourse, and, if he does not concur, he also understands the concept and application of principled disagreement.

Let me also state that, for many reasons, Americans are a justifiably proud people, and it goes without saying that many Americans take great pride in the U.S. Navy. Pride, naturally, is not always a positive thing, however, especially when it is excessive or misplaced. Excessive pride, or hubris, can blind its partisans and lead to overconfidence and jingoism. Jingoism, a more warlike familiar of traditional national pride that is substantiated by a prosperous economy and worldwide interests, was once very much the domain of the British Empire. Now it has found a more affluent and comfortable home in America, the only major industrialized country that was lucky enough not to endure large-scale attacks on its homeland in World War II. This hubris has been recognized by America watchers worldwide. In suggesting that the national disease of the United States is megalomania, Margaret Atwood may well have been right.[4] The United States does seem to suffer "an unnaturally great desire for power and control," as defined by the *Cambridge International Dictionary of English,* and Americans frequently seem to fall victim to "the belief that [they] are very much more important and powerful than [they] really are." If this is true of the United States in general, then is it not natural to suggest that the U.S. Navy may also suffer from this affliction?

After all, a goodly number of our American friends have made, over the past sixty years, many over-the-top statements about the prowess of their navy and their armed forces in general. In recent years, as an example, I have rolled my eyes after seeing young Americans wearing t-shirts proclaiming: "United States Navy: The Sea is Ours." They claim that

U.S. technology is far ahead of all others and that no one is their peer, let alone their superior, on the seas. American presidents and statesmen routinely assert that the U.S. military is "the best trained, the best equipped, the best led." (One retired American admiral recently claimed that American Sailors are also the "best-educated" in the world.)[5] In the 2000 book *The Navy,* Vice Adm. Robert F. Dunn, USN (Ret.), president of the Naval Historical Foundation, proclaimed, "At the dawn of the twenty-first century the United States Navy is second to none in size, quality, readiness, and in the professionalism of its sailors and officers. It trains in, visits, and patrols all the oceans and seas of the world. This ability to maintain a capable worldwide naval presence is what separates the United States Navy from all other navies of the world and is its major contribution to world peace."[6] In 1993, a noted historian, Michael Isenberg, referred to the "Cold War U.S. Navy" as "the master of the world's oceans," adding, "This omnipotent fleet was a mailed fist that wore a velvet glove called 'international peace,'" and "The Cold War Navy was supreme on any part of salt water American policy chose."[7] American admirals have also used words like "invulnerable" and even "invincible" to describe the capabilities of the nation's capital ships at various times.[8] Tall words these are, but one should always be careful about letting one's ego promise something that the body and mind cannot deliver.

In his prize-winning 1997 essay "We Are Not Invincible," Lt. David Adams, USN, lamented the blind arrogance of some U.S. Navy admirals:

> Admiral William Flanagan, Commander-in-Chief, U.S. Atlantic Fleet, recently boasted of the Navy's invincibility: 'I would hate to fight an American right now. You would lose so bad your head would spin.' He also suggested that our technical and information dominance can fulfill the public demand for zero casualties: 'If we have the technological advantages . . . ; if we have control of the information . . . and we control the timeline, why don't we just pitch a shutout? You see the American people have put a standard on us that is good for us. They get zero [casualties]; you get the victory. That's good military thinking.' The Admiral's comments reflect the military's preoccupation with Utopian technical solutions and its ignorance of the political danger of cultivating a strategic mind-set that portends prompt, decisive victory with few—if any—casualties. A similar combination of military delusions and political insularity after World War II contributed to our defeat in Southeast Asia.[9]

It is even more distressing to note that a substantial number of Americans have actually bought into this boosteristic and naïve chop-logic. These folks, unlike their more liberal countrymen, are sometimes quite unabashedly hawkish, and some brag that their grand fleet of supercarriers, cruise missiles, nuclear submarines, and surface ships absolutely and unquestionably rules the seas now as Britannia once did and, more than that, that this fleet is practically unchallengeable. We're the biggest and the best and that's all there is to it. After all, they say, with the former Soviet Navy largely immobile, divided, decaying, deceased, or remaining indefinitely at dockside, who can challenge American naval dominance today?

The U.S. Navy is absolutely the biggest and most expensive navy in the world, that is true, but one should not make the mistake of confusing size and hypothetical striking force with real operational ability. Moreover, if one looks back over time and is objective, emotionally detached, and, most importantly, *intellectually honest,* one can plainly see an embarrassing pattern of failure and underachievement, with pivotal combat climacterics (such as the victory at Midway) resulting mostly from the miscalculations of enemies rather than from any other single factor. The purpose of my disquisition is to describe and elaborate this historical pattern of failure and underachievement (not just the issues facing today's Navy), and then to ask a very pertinent but controversial question: Is the U.S. Navy truly the most capable navy in the world, or is it closer to an overrated paper tiger, one that often fails to learn fully from mistakes, whose dominance can be at least partially attributed to the mistakes of former adversaries and that can be cut down to size rapidly by a determined but asymmetrical foe? This is a touchy subject—I think Margaret Atwood put it best when she said, simply and directly, "You need a certain amount of nerve to be a writer"[10]—but rest assured I will do my best to perform the task at hand with all due respect and sensitivity.

Please also note that this is not so much a comparison test between the U.S. Navy and any or all others as it is a "Let's look at the claims made that these people are absolutely the best and see if we cannot find some examples of their not being so." Thus I am not arguing that the U.S. Navy is, for example, inferior to the Chinese People's Liberation Army Navy or any other, per se, but I do wish to challenge the basic and widespread assumption that American sea power is as singularly dominant or powerful as some people claim. In fact, nothing would make me happier than to see the U.S. Navy in the position it claims already to be in—that of the preeminent naval power of the world, absolutely dominant and largely unbeatable in any and every operational facet. The below, as much as a critique, should be seen as my contribution to the U.S. Navy, my propositions for how it can reach that pinnacle. The service's refusal truly to see its own problems means it also refuses to look for ways to improve. If it is not acknowledged, a problem will never be solved. This is really where the U.S. Navy's dilemma lies, in its inability to face facts and respond to them appropriately. My goal is to point out the problems, the ones that are hidden away and not talked about. By revealing them, I hope to make it worthwhile to consider solutions. This I have done. The solutions are implied throughout the book: learn from the best practices of other navies and see if they can be applied to the U.S. Navy in one way or another. In other words, the U.S. Navy must stop acting like an ostrich. It's that simple. That is the solution, or at least a goodly part of it.

I will begin by discussing various international naval exercises that have pitted the supposedly hegemonic U.S. Navy against foreign diesel-powered attack submarines (SSKs), with many ending with very poor results for the Americans, and how American naval officers have withheld, or been told to withhold, information about exercise defeats, especially those involving aircraft carriers. I will also discuss how the U.S. Navy benefited handsomely from the mistakes of the Germans and the Japanese, plus the antisubmarine (ASW) experience and equipment of the British and Canadians (combined with captured German torpedoes)[11] to buy enough time to establish itself as the dominant naval power at the end of World War II, but one with many subtle and not so subtle weaknesses. I will describe the U.S. Navy's nearly continuous neglect of ASW and how its obsession with supercarriers and nuclear submarines has retarded the combat capability of the surface navy and forced the U.S. Navy to rely on allies for essential services. I will demonstrate through historical case studies how bigger is not better in war and that U.S. naval aviators, even Top Gun graduates, frequently do not measure up to those from various air forces. I will also discuss how racism, overwork, and the unpopularity of the Vietnam War eroded U.S. naval power in the 1970s, which led one admiral to confess that the U.S. Navy would have lost a war against the Soviet Union. I will discuss how drug addiction, a bloated personnel structure, bad promotion policies, and an overweight and poorly educated populace have undermined the fighting skills and capacity of the U.S. Navy and how other, "lesser" navies have done better in some ways.

Throughout, I will provide examples, some based on unscripted exercise scenarios and others drawn from real life, that illustrate the many unfortunate and often ignored (or deliberately concealed) deficiencies of the U.S. Navy. Among other things, it will become painfully apparent that unscripted ("free-play") exercise evolutions strongly suggest, almost ineffaceably, that foreign diesel submarines are quite dangerous to the U.S. Navy and that it needs the help of smaller allies in several key areas of naval warfare. I will also suggest, with all due respect, that there is good reason to believe that the mighty U.S. Navy is simply overrated, frequently unable or unwilling to learn, and in dire need of reform. In doing so, I will present a long list of woes that have afflicted or still afflict the U.S. Navy. These woes should not be viewed in isolation. Other navies have similar problems, and some may be even worse in one way or another; the Soviets/Russians have been and are well known for alcohol and morale problems, for example, and the Japanese, for all their ferocity in battle and iron discipline, have a tradition of thinking inside the box (in other words, creativity, and independent decision making are not their strong suits). Also, perhaps some other navies also withhold information about unfavorable exercise results. The difference, however, is that not every other navy goes out in the world and tells everyone, and instills in its personnel, the notion that it is unbeatable and the master of the seas—all the while ignoring or discounting substantial evidence that suggests that intangibles—top-down management over leadership, micromanagement, a zero-defect mentality, and an obsession with technology—produced by outdated and ineffective personnel practices, policies, regulations, laws, and beliefs diminish the returns upon all the resources poured into the Navy. On top of this, senior leaders refuse to change the service's way of doing things when bad things happen. Again, the back-stabbing, no-holds-barred competition created by the personnel system's individual ratings, centralized selection, and promotion boards create survivors who feel they are the best of the best. Thus, they are blindly devoted to a tainted organization: "It selected me, so what's wrong with it?" The U.S. Navy and the Pentagon seem to be the leaders in this particular realm, and my job here is to call them on it.

Before asking you to consider the documented examples below, I would like first to offer a counter to the most likely argument against my findings.

2

The "Exercises Aren't Real" Argument: My Riposte

THE EXAMPLES BELOW ARE from exercise scenarios. Some will say that one cannot draw conclusions from exercises, because they cannot fully duplicate the reality of combat. Some might also say, erroneously, that exercises are only instructional, or academic, using scripted situations with predictable conditions and rules to train the crews on drills and procedures, not actually how to "fight the ship." According to the *DOD Dictionary of Military and Associated Terms*, a controlled exercise is "an exercise characterized by the imposition of constraints on some or all of the participating units by planning authorities with the principal intention of provoking types of interaction."[1] In this kind of exercise, the crews are basically just practicing their various skills, such as gunnery, antisubmarine warfare (ASW), or damage control, and learning how to operate damaged or degraded systems. In other words, they are *learning* about combat, not engaging in it. In these controlled, unimaginative, scripted exercises, there are not supposed to be any winners or losers, and certainly no one worth his salt calls the media to report a "success" in such exercises. This is just part of the complex exercise equation, and it is not the part that interests me (except in those cases in which the rules, while appearing on paper to be restrictive or unfair, truly reflect the political realities faced by democracies in war or conform to the historical reality that many expensive weapons often do not work as advertised).

Nonetheless, I am engrossed by controlled exercises in which enemy submarines disregard the rules to see if the U.S. Navy is really as good as it claims to be. Such was the case in the September 1998 UNITAS exercise, which involved the U.S. Navy and several South American navies. During the exercise, "enemy" diesel submarines were supposed to keep moving at all times, making them easier targets for American sonar teams. This script was unrealistic, and so "enemy" diesel submarine commanders decided to violate the rules by sitting silently on the bottom, which is quite difficult for nuclear submarines to do. According to reporter Bradley Peniston, this unscheduled and uncalled-for realism irritated the Americans. "Local pride can get in the way of useful practice. Helicopter crewman Harder was eager for the rare opportunity to hunt foreign diesel submarines but found some of the Unitas navies weren't playing by the rules, which insist the subs keep moving. 'It's all pride,' the helicopter sensor operator said. 'If they're on battery sitting on the bottom, I'm not going to get them.'"[2] The American actually complained that his side would have found the diesel submarine and attacked it—*if only the enemy submarine and her devious commander had cooperated!* Harder was vexed that what was supposed to be an unrealistically easy target had a mind of its own, just like a bona fide enemy. This is just like playing darts and expecting the bull's-eye or triple twenty to move about in order to be where your dart impacts, then making a fuss when you find they do not play according to these unrealistic expectations.

I am also very interested in "force-on-force" exercises or evolutions, during otherwise controlled exercises, in which there are indeed victors and the vanquished. In these exercises, which closely simulate combat, no ship, submarine, or aircraft has any special advantage or disadvantage. NATO and the U.S. Department of Defense (DOD) define a "free play" exercise as "an exercise to test the capabilities of forces under simulated contingency and/or wartime conditions, limited only by those artificialities or restrictions required by peacetime safety regulations."[3] The purpose of these evolutions is not to train crews but to fight and, ideally, win. As Robert Coram put it, "In a free-play exercise—no scenario and no rules—the orchestrated performance was tossed out. There is no better way to select and test combat leaders than by free play. Free play means winners and losers; it means postexercise critiques. . . . Careerists hated free-play. . . . True combat leaders loved it."[4] In these evolutions, rival crews do their very best to win, as considerable bragging rights accrue to the winners. Realism is important in these exercises. Exercise Tandem Thrust '99, an unscripted multinational free-play exercise, was "as close to war as we can possibly get," said Cdr. Al Elkins, USN.[5] "We're in this exercise like we're in a hot war. When our aviators take off, they have no idea what kind of threat is coming."[6]

No reasonable person would suggest that a ship that regularly fails in free play or unscripted exercises is nevertheless in good shape for combat, and vice versa. Now assume for just a moment, rather than a list of failures, a detailed list of U.S. Navy *successes* in exercises. Suppose a modern and recently worked-up U.S. Navy destroyer "sunk" a worn-out and unreliable World War II–vintage foreign destroyer with an inexperienced crew in a hypothetical free-play exercise. It would be outrageous for the skipper of the obviously outmatched ship to say, "Yes, but exercises aren't reality. In a real battle, my broken-down old ship and her decrepit guns would have clobbered that new American destroyer and her Tomahawk missiles." That would be preposterous, and so is the claim that free-play exercises, like the ones described below, are inherently meaningless. The fact is that *consistent* unscripted exercise results (successes or failures) are useful, meaningful, and provide reasonable analytical tools. And if free-play exercises are not meaningful, why does the U.S. Navy invest so much time and money in them? It does so because these types of exercises frequently reveal both the good and bad news about how a navy might fare in a real war. Consider the many interesting quotes one gets when googling "purpose naval exercise." I did not see a single "just having a good time" and "shooting the breeze" statement. While not always the case, the standard, antediluvian excuse employed by the Navy's apologists that *all* defeats (even in free play or unscripted exercise evolutions) are *purely* because the U.S. ships or aircraft involved were operating under some sort of artificial restriction, unrealistic limitation, or handicap is also often rather spurious, exaggerated, overly convenient, deceitful, or just a cop-out, and I will deal with that matter in due course.

I also do not fully accept the whole "These exercise defeats only involved allied forces, and our allies are much better than our potential rivals, so it doesn't matter" argument, either, for the United States certainly has a long tradition of underestimating its enemies—Japan (initially at least), North Korea, China, and North Vietnam come to mind—and besides, if a friend driving a quiet diesel submarine can sink a carrier or nuclear submarine, what's to prevent a rival from developing the same skills to do so? Tom Clancy claimed in 1999 that the only navies in the world that could possibly do any harm to the U.S. Navy are those of current allies, and being friends, they "are not likely to do that."[7] As we will see in the pages that follow, he was wrong then, and he is wrong now, because courage, motivation, training, leadership, and professionalism are not proprietary resources owned and trademarked by the Western countries, and to say otherwise is pure ethnocentrism—a more polite version of a rather nasty word that many Chinese readers have already used to describe some of Clancy's stories. (Indeed, Clancy must have been surprised when *The Washington Times* revealed in 2006 that a Chinese diesel-electric submarine had successfully penetrated the screen surrounding the USS *Kitty Hawk*, undetected).[8] The technology can be purchased from any number of countries, and the skills can be developed by any nation with the political will to do so, be it big or small, rich or poor, friend or foe. I suggest one never forget that "the fateful attack on Pearl Harbor forced the Western world to revise its opinion of Japan's airmen. Before the war, Japanese aviators had been seen as figures of ridicule and disdain; yet the ruthless skill and efficiency of their performance in December 1941 and the months that followed won them a new reputation as a breed of oriental supermen."[9] As Col. Thomas Hammes, USMC, has said, "Any nation that assumes it is inherently superior to another is setting itself up for disaster."[10]

On yet another level, some will also claim that the fact that exercises are conducted in relatively small areas makes it easier for diesel submarines to detect and attack surface ships. In real life, they assert, the oceans are much bigger and it is more difficult for a diesel submarine to position itself to attack a much faster carrier battle group. I would ask those who support this argument to consider two things.

Firstly, many U.S. surface combatant ships were sunk in the open ocean by slow, primitive diesel submarines in World War II, including the carriers USS *Yorktown*, and *Wasp*, the escort carriers *Liscombe Bay*, and *Block Island*, the cruisers *Indianapolis* and *Juneau*, the destroyers *Mason*, *Reuben James*, *Satterlee*, *Jacob Jones*, *Hammann*, *O'Brien*, *Porter*, *Henley*, *Buck*, *Bristol*, *Leary*, *Leopold*, *Fechteler*, *Fiske*, *Eisele*, *Shelton*, *Eversole*, and *Frederick C. Davis*, and many other types of surface ships.[11] U.S. battleships were damaged by submarine attacks and taken out of action for long periods of time as well. In one case the 35,000-ton battleship USS *North Carolina*, one of the most powerful and up-to-date ships of her time and far more advanced than the ships destroyed at Pearl Harbor, was taken out of action for two months *by a single torpedo* fired by the Imperial Japanese Navy's submarine *I-19*.[12] The carrier *Saratoga*, which was "the largest warship in the world"[13] when she was launched, "was torpedoed on two separate occasions early in the war and was out of service for months."[14] In one battle, a single torpedo from a Japanese submarine left the 33,000-ton carrier "dead in the water" for several hours; she had to be taken under tow by a cruiser.[15] In addition, the Imperial Japanese Navy's 71,890-ton supercarrier *Shinano* was sunk by a diesel submarine, as was the 36,000-ton fast battleship *Kongo*. Submarines also claimed several British carriers.

Secondly, consider that even though carriers and surface ships are more advanced today and are still much faster than conventional submarines, that does not give them any additional life insurance, because in a war the enemy diesel submarine will know (a) where the U.S. Navy ships are coming from and (b) where they are likely headed. They do not have to catch up to a carrier battle group making more than thirty knots; they can just wait for it, and no one can predict exactly where. The only protection the U.S. Navy will have is solid ASW skills, and as we will see in this analysis, the assumption that it has such skills is not well founded. Today's diesel submarines are far better than those of the past, and with the U.S. Navy now concentrating more on the dangerous, noisy, and shallow waters of the littorals, if anything, the potential threat from quiet conventional submarines is greater now than it was in World War II.

One more thing about exercises. I have noted over the years that our U.S. Navy colleagues expect to always win, by virtue of what they earnestly believe is superior technology (on which some say the U.S. Navy has grown overly dependent, and consequently, rather sloppy) and/or superior training. They simply cannot fathom the results when things do not go their way all the time. When a real crackerjack American F-18 squadron beats a foreign squadron in air combat maneuvering (ACM), for example, the U.S. Navy's supporters do not ask questions about exercise parameters. They just assume that American technology and training were better, so case closed. However, when a U.S. ship or squadron loses in a competitive free-play or unscripted exercise, the response is rarely, "Well, you can't win them all" or "You win some, you lose some." Sadly, the more typical response is to call a foul at the very concept of being beaten. Were the conditions unfavorable to the U.S. Navy? Did the U.S. Navy fighters lose because they had to carry more fuel tanks and were therefore less agile, or had fewer landing bases available, than their land-based opponents?[16] Was the exercise unfair to U.S. forces (as if war could ever be "fair")? Remember former President George H. W. Bush, a Navy veteran, who said the following after a U.S. ship shot down an Iranian airliner: "I will never apologize for the United States of America. I don't care what the facts are." I find this quote very much in keeping with the *nihil ad rem* culture of evasion, excuse making, obfuscation, blame shifting, buck passing, and denial in the U.S. Navy, and I urge you to keep this in mind as you read this book. Denial, in the words of military commentator Stan Goff, is indeed "the grandest of American appetites."[17] The late Col. David Hackworth, USA (Ret.), was indeed justified in saying, "The Navy is afflicted with arrogance and self-denial,"[18] and this evaluation will amplify that point.

Of course, the U.S. Navy is not the only navy that wishes to avoid negative exercise results. For example, the pre–Pearl Harbor Japanese naval war games were clearly rigged to make it easier for the Japanese to win. Nevertheless, said Gordon Prange, "Even with the umpires bending over backward in favor of the home team, *all had agreed that they must anticipate the sinking of several carriers*" (emphasis mine).[19] This, I would suggest, is probably more intellectually honest than many American examples, especially under the tenure of John Lehman, as we will see in a few moments. I would argue that the tendency to make aircraft carriers officially "indestructible" is, if anything, especially strong in the U.S. Navy, because its supercarriers are, by several orders of magnitude, the most expensive naval ships in the world and thus they are tempting targets for budget cutters and disgruntled taxpayers in a modern democracy. Secondly, the U.S. Navy's rather harsh "up or out" promotion system ensures that careerism, and the consequent desire to have a "perfect record," encourages intellectual dishonesty and an unwillingness to acknowledge poor performance. I will return to this matter later, too.

As for methodology, the first section of the book relies on qualitative rather than quantitative data. The reason for this is simple. As one reservist, Capt. Dean Knuth, USNR (Ret.), will attest later, the U.S. Navy keeps a tight lock on its exercise evaluation data, especially potentially embarrassing failures. These exercise reports, note well, are not available to the general public, and attempts to make them public have been rebuffed by the Navy. Under these conditions, a statistical analysis is not likely. In fact, after conducting a thorough search of the available unclassified materials, I could not locate even one such study, and one can be sure that is just what the U.S. Navy wants. This study is meant to be the opening of a discussion, and thus my purpose is merely to ask questions and raise issues, rather than to answer all of them comprehensively. My task here is to try to put the pieces together and see if any conclusions can be supported or extrapolated. Although they would be helpful, one does not always need reams of statistical data and tables to recognize a plain fact, especially when history, common sense, and credible authorities support the conclusion. We do not require a statistical analysis to understand universal truths.

I always liked the way Bruce Russett lucubrated his methodology, so I shall indicate my concurrence by quoting him directly: "My intention is to be provocative. . . . The argument is not one subject to the principles of measurement and the strict canons of hypothesis testing—the mode of inquiry with which I feel most comfortable. Nevertheless the subject is too important to leave untouched simply because the whole battery of modern social science cannot be brought to bear on it."[20] Like Fallows, my wish here is to be "suggestive, rather than encyclopedic or definitive," and as was the case with Fallows' 1981 magnum opus *National Defense*, "Much of the story is told through anecdotage and case history, but these particulars are meant to suggest certain casts of mind, certain rules of organizational life."[21]

I would also add that it does not require a leap of faith to know that there is no such thing as an unsinkable ship, no matter how big it is, how many watertight compartments it has, or how much armor plating it has. Nor does it require much imagination to comprehend that a nearly silent diesel submarine can most definitely stalk and sink even the largest surface warships (or, these days, nuclear submarines) with relative ease. Diesel submarines are not and have not been necessarily restricted to home or coastal waters, either, contrary to what many nuclear submarine advocates emphasize. In fact, many diesel submarines have been "forward deployed" thousands of miles from their home bases and have operated against the enemy on the other side of the ocean. Such things happened in both world wars and during the Falkland Islands war of 1982, and they can happen today. Even Compton-Hall, whose writings evidence a pro-nuclear submarine slant, once cautioned, "It is a great mistake to denigrate SSKs: they will continue to be a menace for the foreseeable future and the Soviet Navy knows it."[22] Those who deny these facts are in fact denying reality. As Aldous Huxley once said, "Facts do not cease to exist because they are ignored."[23]

3

David *vs.* Goliath: Diesel Subs and Mines Take On the U.S. Navy

> Even in the open ocean NATO fleet exercises demonstrate, time and again, that a proportion of SSKs will get through the screen.
>
> —CDR. RICHARD COMPTON-HALL, RN (RET.)[1]

> U.S. Navy exercises with diesel submarines since the mid-1990s have often proved humbling.
>
> —JOHN BENEDICT, NATIONAL SECURITY ANALYSIS DEPT., JOHNS HOPKINS UNIVERSITY APPLIED PHYSICS LABORATORY, 2005[2]

> There are a lot of them (diesel submarines), and they're hard to detect.... especially in local waters where they know all the tricks. One of them could sink a carrier easily, and they cause a lot of anxiety.
>
> —CDR. TOM FITZGERALD, USN, 2006[3]

IN 1952, THE FIRST MAJOR NATO naval exercise, Operation "Mainbrace," was conducted in the North Atlantic. Involving eighty-five warships from the United States and the United Kingdom, the exercise was the brainchild of none other than Gen. Dwight Eisenhower, who wanted to demonstrate to the satisfaction of Norway and Denmark that NATO could indeed protect them in the event of a Soviet attack. Three U.S. Navy carriers participated (the USS *Midway, Wasp,* and *Franklin D. Roosevelt*).

The captain of the *Roosevelt* encouraged his crew to be vigilant in the face of a significant diesel submarine threat. Said Cdr. George W. Anderson, USN, "Any man who spots a periscope before it attacks gets special liberty to London."[4] Anderson's crew soon got their chance to deal with a sneaky diesel submarine, HMS *Taciturn,* when the boat reportedly "got through the destroyer screen and promptly claimed hits"[5] on all three U.S. carriers, and other ships, with conventional torpedoes.

Curiously, although nuclear weapons were available at the time, simulation of their use was not included in the exercise scenario. Why not? Hayes, Zarsky, and Bello have said that the Navy had convinced itself that it could "survive" nuclear warfare,[6] based on the allegedly favorable results of the atomic tests during Operation Crossroads at Bikini atoll in 1946, but many do not see it that way at all. According to Mike Moore, "The ostensible purpose was to evaluate the effects of nuclear weapons on modern naval vessels so that design improvements could be made. In fact, Crossroads was mainly a public relations exercise" that "became at times a circus, a farce, a near tragedy."[7] "Of the 97 vessels, 22 were sunk outright. The 73 that remained afloat, such as the battleship *New York,* were hopelessly irradiated," noted James Delgado.[8] Despite repeated warnings from atomic experts that radiation would kill the crews even though many of the ships were still intact, the Navy did not want to hear this and did not consider the contaminated ships to be true "casualties." And keep in mind that even though no one knew its possible effects at the time, the electromagnetic pulse produced by a nuclear explosion close by would surely have fried the electronics on all the ships as well. This might explain why the Navy did not wish to simulate the use of atomic weapons during Operation Mainbrace; it did not want this issue to raise its ugly head again, or get into the newspapers.

When the *Taciturn* claimed its victories, the exercise umpires, all on the surface ships, did not concur; they initially ruled that the submarine herself had been sunk. The matter as to "who got whom first" was supposedly subjected to a postexercise review, but the definitive answer was, to my knowledge, never made public. Although in this case it was never proven that the submarine had been successful, at least not publicly, it is not at all far-fetched for a single diesel submarine to attack three major surface ships successfully. That very thing happened in World War I, as Richard Compton-Hall once described, when a "pathetic" German submarine, the *U-9,* took on and destroyed three British cruisers in one day. It is also not at all far-fetched that U.S. Navy officers might overlook, elide, or fail to intromit

DAVID VS. GOLIATH

Despite all the denials, U.S. aircraft carriers, even when they have numerous escorts, have proven to be quite vulnerable to diesel submarines. —T. Cichonowicz, U.S. Navy photo

successful attacks against the aircraft carriers that have formed the very basis for U.S. naval power projection over the past sixty years.

There have been many other exercises in the years since, but only a handful of these have become public knowledge, usually in the pages of a few periodicals and base newspapers. One exercise that did draw public attention was in 1973. The exercise was code-named "Uptide," and during it, according to Thomas B. Allen, the aircraft carrier USS *Ticonderoga* (which has since been retired, its name now inherited by a cruiser) was sunk twice by enemy submarines and taken "out of action."[9] This defeat, however, remained officially unreported and strictly "off the record." Later, in 1981, the NATO exercise Ocean Venture ended much the same way for the U.S. Navy, with submarines destroying carriers, but this time something very different and controversial happened—an exercise analyst had the audacity to try to report the truth, and he paid for it later.

Before I get to those ugly details here is a little background information from the exercise senior analyst, Lt. Cdr. Dean Knuth, USN:

> In September 1981, the largest exercise in Atlantic Fleet history reached a peak after a two-carrier battle group completed a transit across the Atlantic. The ships entered the Norwegian Sea and their planes struck

simulated enemy positions in waves of coordinated air attacks. The NATO exercise was Ocean Venture/Magic Sword North, and it was the first time that the Commander in Chief, Atlantic Fleet, had amassed two American aircraft carriers, the British through-deck cruiser *Invincible*, and a large supporting force which included Royal Navy, Canadian Navy, and U.S. Coast Guard ships—all for the purpose of demonstrating the ability of the free world 'to control the Norwegian Sea and contain Soviet sea power.[10]

During the exercise, a Canadian submarine slipped quietly through a U.S. Navy aircraft carrier destroyer screen and conducted a devastating simulated torpedo attack on the carrier. The submarine was never detected. A second carrier was also reportedly destroyed by another enemy submarine.

Knuth tried to use material from his official report in a magazine article, but when Navy officials read a draft of it, his work was promptly censored to minimize the potential fallout. Some might argue that the Navy had good reason to do this, as ostensibly a matter of "national security," but I find that claim a bit of a reach, because *everyone* knows that diesel submarines sank big aircraft carriers and other major combatant ships in World War II, as I have already mentioned, and there have been no great breakthroughs in surface ship survivability since then. The article was never published. Said Knuth in a subsequent newspaper interview, "The fact is our aircraft carriers were successfully attacked by torpedoes or missiles from submarines in our major exercises."[11] This would tend to contradict former Navy secretary Charles Thomas's boast that U.S. Navy carrier task forces are "the toughest target in the world."[12]

In 2005, Captain Knuth, USNR (Ret.), told me that "We were interfered with by upper echelons of the Navy who wanted us to delete all references to sub attacks against carriers."[13] According to Knuth, Secretary of the Navy John Lehman was trying to convince Congress to fund two new additional aircraft carriers, and his case could have been seriously undermined if Knuth's original manuscript had come into the public eye. In Ocean Venture '81, "90 percent of the first strikes were by submarines against the carriers,"[14] and this fact did not sit well with many naval aviators, or with Lehman. In fact, Lehman resorted to ad hominem circumstantial attacks and cheap shots against Knuth in the media, dismissing him as merely a "retired Lieutenant Commander"—even though Knuth was still serving on active duty. As we all know, such tactics are commonly used when someone does not like hearing the truth—they simply bypass the opposing argument altogether and attack the person making it. At that point Knuth got "fed up with the politics" of the regular Navy and transferred to the Naval Reserve, where he was eventually promoted all the way to captain and became the commodore of Naval Coastal Warfare Group 2 (Atlantic). Had he stayed in the regular Navy, Knuth doubts that he would ever have gotten another promotion, let alone two. He became persona non grata in the regular Navy.[15]

This ties in with the Navy's need to retain what Michael Isenberg calls the "illusion of American omnipotence": "Always, the primary goal was to retain public confidence in the Navy, no matter how silly or god-awful the blunders."[16] This is necessary when one runs the most expensive fleet in the world, in a democracy in which the government is supposed to be accountable to the taxpayers, and especially if one is too stubborn to ever change one's ways. "Like everyone, the Navy loved good news, but the institution loathed bad news like the very devil. Bad news could lose votes, lose money, lose public support. As a result, and without a doubt, the Navy was in the business of managing its news. . . . The best Navy PR people told the truth, or as much as they were allowed to tell. A few others other, however, produced merely the advertising garbage of Madison Avenue, dealing mostly in evasion, half-truths, propaganda, and lies."[17]

Although the Navy tried to hush the matter up and ordered Knuth to destroy his original manuscript, he kept a copy of the censored version; even in its expurgated form it is interesting and titillating reading. In the censored version, titled "Lessons of Ocean Venture '81," Knuth expatiates that the carriers *Eisenhower* and *Forrestal* "would never have made it to Norway in a wartime situation" because of the submarine threat.[18] He continued:

> The first major event of the exercise was strictly a World War II leftover not likely to take place in the future: carrier against carrier. The *Forrestal's* battle group steamed in total emission control and sneaked toward the *Eisenhower* group which was on track for the Greenland-Iceland-United Kingdom gaps. This event was parochialism personified. In Battle of Midway style, the aviator admiral relied on long-range tactical air strikes against the *Forrestal*, with little or no fighter air support. The surface admiral dispersed all of his surface combatants away from his carrier and sent them quite effectively on an antisurface mission against the

Eisenhower. Unfortunately, in doing so, he unrealistically left his own carrier open for submarine and air attack.[19]

Knuth continued, "The most exciting part of the exercise was the transit of the Iceland–United Kingdom gap. In the previous five autumn NATO exercises, the carriers have always been attacked going through the gaps."[20] It is often said that in war the first casualty is truth, but in this case I would say the first naval casualty in a general war with the Soviet Union would have been the *lie* that U.S. Navy aircraft carriers are invulnerable. Fallows made the same argument in 1981, saying those big ships would be the first to go down if things got nasty.[21]

The USS *Eisenhower* was successfully attacked by a surface ship, said Knuth, but official reports by the commanders on scene seem to have overlooked this success:

> An Orange [that is, "enemy"] missile ship sneaked to within weapon-firing range during the night and maintained station on the *Eisenhower*. At sunrise, the ship simulated emptying her missile load into 'Ike' without herself being engaged until after signaling that she was engaging the carrier. The surprise attack was well described in traffic among warfare commanders on the satellite circuit, but when the carrier striking force summary report was received by the fleet commander, it stated that the Orange ship had been tracked and that a Blue ship, stationed between the carrier and the Orange ships, had been watching his actions. The report described a far different action than the confusion that had existed at the time of the engagement.[22]

There was also an apparent "friendly fire" incident in which "a guided missile destroyer in Ocean Venture mistakenly harpooned the *Eisenhower*, mistaking a carrier for an Orange surface combatant. The composite warfare commander was so furious that he threatened to excommunicate the ship from the battle group."[23]

Knuth was remarkably sedulous in offering thorough criticism of U.S. Navy battle group tactics, organization, intranavy parochialism (aviators versus surface warfare and submarine rivalries) but spoke very highly of the British contingent: "The British force employment, asset management, commands and action reports were superlative and a model for our battle group to emulate."[24] He also conceded that British officers and men "are better trained than our best and their battle group commanders and staffs are highly proficient in tactics. My professional note in the December 1981 *Naval Institute Proceedings* explains in depth why this is the case."[25] Finally, Knuth lamented the fact that "our battle groups continually prostrate themselves before the hard-to-find enemy because of our perception of our own invulnerability.... The enemy can locate battle groups easily, and with a large fleet of submarines, set up for a pre-planned attack. Our policy is normally to head straight for danger and not shoot until shot at first. When the Orange force makes a preemptive attack, it is usually of such a magnitude that the battle group is overwhelmed and lost."[26]

Despite the Navy's censorship of the Ocean Venture '81 article and the fact that the redacted version was never published, the story became public knowledge in Canada. An anonymous Canadian submariner leaked the story to a Halifax newspaper and indicated that this successful Canadian attack on an American carrier was by no means an isolated incident. It had been a simple ambush in the North Atlantic, and it had worked perfectly. Indeed, the article concluded that the Americans never knew what hit them, that they were embarrassed by this failure, and that they wanted to bury the matter then and there. The Canadian submarine did not fire the customary green flare to indicate a hit, for reasons unknown to anyone except for the skipper of the submarine, but instead simply took periscope photos of the carrier to prove its point. In doing so, the diesel submarine ambushed a surface ship in the same way that Germany's U-boats had decades before. This news and Knuth's original uncensored report, which ended up in the hands of Senator Gary Hart, caused quite a stir in Congress, and the U.S. Navy had a lot of explaining to do. Why had not one but two American carriers been sunk, and why had the submarines responsible not been detected? Why indeed had a small, 1960s-vintage diesel submarine of the underfunded and multidimensionally "bantam" Canadian navy been able to defeat one of America's most powerful and expensive warships, and with such apparent ease? And let us not forget that in the 1980s, a U.S. carrier battle group could include perhaps eight surface escort ships, including cruisers, destroyers, frigates, and a nuclear submarine, to provide air-defense and ASW capabilities.[27] Now, unfortunately, there tend to be far fewer escorts. Carrier Group 5, centered on the USS *Kitty Hawk*, for example, currently has only three escort ships, two cruisers and a destroyer, and sometimes the *Kitty Hawk* goes to sea with only two of those escorts.[28]

Conjointly, why were the Canadians able to do essentially the same thing to the U.S. Navy in subsequent exercises in the spring of 1983? The *Winnipeg Free Press* reported that the submarine HMCS *Okanagan* "snuck to within a kilometer of the USS *John F Kennedy,* went through preparations to fire a salvo of torpedoes and slipped away unnoticed by the carrier or the destroyers." The submarine got close enough "to score a lethal hit, Defence Minister Jean Jacques Blais said." Blais went on to say, "This is a matter of some pride for submariners and shows the strength of our underwater boats at a time when satellite detection can identify surface ships more readily."[29]

There are several possible explanations. Firstly, the Canadian submariners have a long-standing reputation for being well trained and professional. Supporting this argument is Compton-Hall, one of the world's leading authorities on submarines, who evaluated the Canadian submariners as "first class, aggressive and innovative."[30] Secondly, the *Oberon*-class submarines then used by the British, Canadian, Australian, Chilean, and Brazilian navies, built in the United Kingdom but based on a German design from World War II, were probably the quietest in the world at that time. Of course, adverse acoustical conditions produced by temperature variations (thermal layers) and other factors may temporarily cloak even the noisiest submarines, but the nearly silent *Oberon*-class diesel boats running on batteries were even harder to find in such conditions. And in any case, Knuth described the acoustical conditions as having been "excellent" for detecting submarines,[31] so the answer probably lies elsewhere. A third possible reason is perhaps that the powerhouse U.S. Navy just is not very good at hunting submarines, especially the ultra-quiet diesel boats available today. It is the last explanation that intrigues me, and it is the one on which I shall focus much of this review.

While Canadian submarines have routinely taken on American carriers, other small navies have enjoyed similar victories. The Royal Netherlands Navy, with its small force of extremely quiet diesel submarines, has made the U.S. Navy eat the proverbial humble pie on more than one occasion. In 1989, naval analyst Norman Polmar wrote in *Naval Forces* that during NATO's exercise Northern Star, "the Dutch submarine '*Zwaardvis*' was the only orange (enemy) submarine to successfully stalk and sink a blue (allied) aircraft carrier." The carrier in question might have been the USS *America,* as it was a participant in this exercise.[32] Ten years later there were reports that the Dutch submarine *Walrus* had been even more successful in the exercise JTFEX/TMDI 99. "During this exercise the *Walrus* penetrates the U.S. screen and 'sinks' many ships, including the U.S. aircraft carrier *Theodore Roosevelt* (CVN-71). The submarine launches two attacks and manages to sneak away. To celebrate the sinking the crew designed a special T-shirt." Fittingly, the T-shirt depicted the *Theodore Roosevelt* impaled on the tusks of a walrus. It was also reported that the *Walrus* sank many of the *Roosevelt*'s escorts, including the nuclear submarine USS *Boise*, a cruiser, several destroyers and frigates, and the command ship USS *Mount Whitney*. The *Walrus* herself survived the exercise with no damage. Talented and wily enemies, of course, usually do not play by the rules, and they do not stick to a script.[33]

Truthfully, it should come as no great eye-opener that Dutch submarines would do well against the U.S. Navy. The Dutch submarine service has an enviable reputation and has been praised by such people as the late Vice Adm. Charles A. Lockwood Jr., USN, who was Commander, Submarines Pacific during World War II. Lockwood said in 1945 that Dutch submarines in the Pacific were "thoroughly effective. They handled their boats with great skill and do not need to take off their hats to anyone." The admiral also mentioned his "high regard for their ruggedness and fighting skills."[34] Nowadays, many navies, including the U.S. Navy, send their submarine officers to the Netherlands to undergo the legendary Netherlands Submarine Command Course. In November 2002, the Royal Australian Navy's official newspaper described the Dutch course for prospective diesel submarine commanders as arguably "the best submarine training in the world."[35] U.S. Navy students who have taken the course have also found it extremely challenging. In 2002, naval officers from the United States, Australia, Canada, Israel, and the Netherlands took the course, but unfortunately, the American officer failed, due to a safety violation.[36] The American officer was the only student to fail that year, but in fairness, he was a nuclear submariner, naturally, and ergo was much less familiar than the other students with the workings of a diesel submarine and its battery operations.

Reassuringly, Lt. Cdr. Todd Cloutier, USN, did graduate from the Dutch course in 2003, and he too confirmed the program's "legendary reputation," describing it as "perhaps some of some of the toughest training a submariner can get."[37] Although this course is for experienced officers who wish to command a diesel submarine, he was very impressed by the overall training received by Dutch junior officers. "A Dutch Junior Officer (JO) with three years at sea is quite proficient with the periscope. During my familiarization ride on *Bruinvis,* I saw a non-qualified JO take the conn and conduct a task-group penetration against a multinational task force. It wasn't perfect, but quite impressive for a JO with less than two years on board."[38] This suggests that a U.S. naval officer of comparable rank would have been less capable.

The preceding section concerned aircraft carriers and surface ships only, but the U.S. Navy has long maintained that its nuclear submarines are clearly and unambiguously superior to any and all diesel submarines. This dogma has been perpetuated for decades, said Rear Adm. Corwin Mendenhall, USN (Ret.), in 1995, because the nuclear submarine force leadership "has been brainwashed by the Rickover nuclear-only philosophy. Consequently, no one in the Navy knows—or wants to know—what diesel submarines' capabilities are."[39] Nuclear submarines are so superior, allegedly, that some U.S. submariners have long said that they need not even worry about conventional submarines. In a 1998 report, Ivan Eland cited an article in which "one U.S. submarine commander reported that he would not even bother to destroy a diesel because he could detect the boat before it detected him; he said that he would simply avoid it."[40] Although this oblivious and ignorant thinking has finally begun to change, there is still much that needs to be done. What follows is intended to challenge that old establishment nonsense, and I hope contribute, in a small way, to its reform.

As we saw earlier, the Canadians have caused serious trouble for U.S. carriers and surface ships, but they have given U.S. nuclear submarines a run for their money as well. Cdr. Peter Kavanagh, Canadian Forces (Ret.), was the skipper of the submarine HMCS *Onondaga,* and he recalls that "In 1996, the sub was 30 years old and they sent us against a brand new American nuclear submarine and we beat the pants off them. So we knew what we were doing. . . . We had to be smarter, because we didn't have the endurance and the speed of the nuclear submarine." Also, "The crew was so well-trained at that point, because we had been together two years. It just goes to show you just what you can do if you're trained well."[41] Since Canada only had three submarines at that time, it is not hard to see how it could maintain a crew continuously for two years or more, and cohesion is indeed an important factor in combat readiness. On the other hand, nuclear submariner Dr. Andy Karam noted that during his tour of the submarine USS *Plunger* that the average annual turnover was about 25 percent, which means that during the same two-year period 50 percent of the crew of a U.S. submarine would leave the ship. This in turn suggests that Canadian submarine crews are better off because they work and train together for longer periods.

Commander Kavanagh gave me the specifics on this exercise in March 2006:

> That particular exercise occurred in Jan 1996 and was against the USS *Hartford*. She is a 688(I) *Los Angeles* Class nuclear attack boat and had just

The USS *Hartford* (SSN 768) was clobbered by a thirty-year-old Canadian submarine during exercises in the late 1990s. —*General Dynamics, Electric Boat Division*

become operational after build (i.e., completed the entire technical and operational trial, training and workup process). The exercise took place in the northern part of the Virginia exercise areas. I don't recall the specific water column conditions, but at that time of year and in that area there was certainly a "layer" that we exploited. We also had an experimental narrowband/broadband sonar suite that gave us excellent detection capability. We had 7 encounters and *Onondaga* bounced *Hartford* 6 times to *Hartford's* one. And I deserved getting smacked on that occasion because I got a bit complacent with a snorting cycle [snorkeling, i.e., periodically running the diesel at periscope depth to recharge batteries]. However, 6 to 1 isn't bad! The *Hartford* of course was an extremely quiet boat and had excellent sonars. We won the day through the use of tactics. I recall hiding in the middle of a fishing vessel fleet, using the water column aggressively, going ultra quiet and slow when I thought he was looking for Doppler, etc. The Americans were stunned with the results—this was their newest submarine after all. We carried an American rider who requested we emphasize our new sonars in the after action report.

Kavanagh noted that he had great respect for the skipper of the *Hartford* and that he "knew his business" well, despite the defeats.[42]

The penumbral Australian submarine force has also scored many goals against U.S. Navy carriers, and nuclear submarines as well. Back in the 1960s and 1970s, for starters, Australian *Oberon*-class submarines were a gigantic pain in the side of the U.S. Navy, as "the ability of the O-boats to run in total silence enabled Australian submarines to successfully attack USS *Enterprise* in a training exercise, despite a huge number of supporting ships 'protecting' it."[43]

On September 24, 2003, the Australian newspaper *The Age* reported that Australia's *Collins*-class diesel submarines had taught the Americans a few lessons during multinational exercises. By the end of the exercises, Australian submarines had destroyed two U.S. Navy nuclear attack submarines and an aircraft carrier. For the Australians, all three ships were easy targets. According to the article: "The Americans were wide-eyed,' Commodore Deeks (Commander of the RAN Submarine Group) said. 'They realized that another navy knows how to operate submarines. . . . They went away very impressed.'"[44] In another statement attributed to Deeks, he declared, "We surprise them and they learn a lot about different ways of operating submarines. . . . The Americans pour billions into their subs but we are better at practical applications."[45]

However, officially, the U.S. Navy, a true military opsimath, soon went into damage control mode and oppugned the idea that the Australians could beat an American nuclear boat in a fair fight. Said the *Bulletin of the Atomic Scientists*:

> The United States is justly proud of its military prowess, but apparently a little defensive when anyone else shows a bit of talent. *Defense Week*'s 'Daily Update' on October 1, 2003, reported that the commander of the U.S. Pacific Fleet was trying to downplay the fact that an Australian diesel-electric submarine had 'sunk' an American submarine during recent training exercises, and said the Australians were making too much of the simulated hit. Adm. Walter Doran said that the outcome 'certainly does not mean that the *Collins-* class submarine in a one-on-one situation is going to defeat our *Los Angeles*–class or our nuclear submarines.'[46]

But if the American submarine was "supposed" to be sunk, using a noise augmenter to simulate a Soviet sub, purposefully running with "degraded" antisubmarine warfare systems (and there is no available

The Australians have "sunk" many U.S. ships, including the USS *Enterprise* (CVN 65). —*L. Wilson, U.S. Navy photo*

evidence to support any of these excuses), why did an experienced Australian submariner like Commodore Deeks, an officer in one of the most professional navies in the world, make what would then have been unsubstantiated, out-of-context, and unfair statements to the media? As Compton-Hall said, the Australian submarine service is "outstandingly efficient,"[47] and it has an excellent reputation.

Now, some might say that Commodore Deeks said this merely to market or hype the RAN's newly built submarines to the Australian public, after a long and troublesome teething period. I do not think that would have been necessary by 2003, however, because the *Collins*-class boats had *already* proven themselves rather convincingly by then, as we will see in a moment. I would wager that, like the Japanese at Pearl Harbor, the Australians had actually caught the Americans off guard and unawares, and the U.S. Navy had to go on the defensive. As we will see later, Capt. Richard Marcinko, USN, strayed from the rules during exercises in the 1980s, and he achieved incredible results. War, as they say, is not fair, and anyone familiar with polemics knows that preemptive or surprise attacks have often proven devastatingly effective, as the Israelis demonstrated in 1967.

In October 2002, the Australians reported that their diesel submarine HMAS *Sheehan* had successfully "hunted down and killed" the nuclear submarine USS *Olympia* during exercises near Hawaii. The commander of the *Sheehan* observed that the larger American nuclear boat's greater speed and acceleration had been no advantage, because "it just means you make more noise when you go faster."[48] In the previous year, during Operation Tandem Thrust, analyst Derek Woolner asserted that HMAS *Waller* sank "two American amphibious assault ships in waters of between 70–80 metres depth, barely more than the length of the submarine itself. The *Collins*-class was described by Vice-Admiral James Metzger, Commander, U.S. Seventh Fleet as 'a very capable and quiet submarine.'"[49] Although the *Waller* was herself sunk during the exercise, the loss of a single diesel submarine, in exchange for two massive amphibious assault ships, is quite a good bargain and very cost-effective.

Finally, during RIMPAC 2000 it was disclosed that HMAS *Waller* had sunk two American nuclear submarines and gotten dangerously close to the carrier USS *Abraham Lincoln*. Even more ominous, asserted researcher Maryanne Kelton, is that "even though the exercises were planned and the U.S. group knew that *Waller* was in the designated target area, they were still unable to locate it. New Minister for Defence, Robert Hill, recorded later that the 'Americans are finding them exceptional boats . . . in exercises with the Americans they astound the Americans in terms of their capability, their speed, their agility, their loitering capacity, they can do all sorts of things that the American submarines can't do as well.'"[50] In 2003, Cdr. Peter Miller, USN, spoke about his experiences with the Australian diesel submarines, and he paid the greatest (politically correct) compliment that a nuclear submariner can make. He said that the Australian diesel submarine was "on a par" with U.S. nuclear submarines and that "the Collins are great submarines."[51]

USS *Abraham Lincoln* (CVN 72), followed by three escorts, was stalked by an Australian diesel submarine in 2000. The submarine got "dangerously close" to the carrier. —*S. Call, U.S. Navy photo*

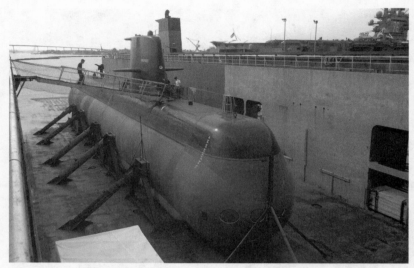

Conventional submarines, like the Swedes' HMS *Gotland*, have made life very difficult for U.S. Navy antisubmarine warfare units. — *J. Sims, U.S. Navy photo*

The *Collins*-class submarines were designed in Sweden, and naturally the Swedes themselves have been able to raise some eyebrows in the U.S. Navy. In a 2004 presentation in Stockholm, Vice Adm. Kirkland H. Donald, USN, affirmed that "Today, Sweden manufactures some of the best built and equipped submarines and surface ships in the world. The *Gotland* class is not only quiet, but has a most impressive combat system. If I remember correctly, in the fall of 2000, there was a multilateral, blue-water ASW exercise where Commander Gumar Wieslander and his crew in HSwMS *Halland* demonstrated remarkable prowess exercising against one of our finest ships, USS *Annapolis*. That exercise, along with many others, reinforced the difficulties in prosecuting a well built, well maintained diesel submarine, with a well trained crew."[52] He did not say that the Swedish boat "sank" the *Annapolis*, but the subtle implication might be there if one reads carefully between the lines.

The Japanese have also proven to be formidable in their modern diesel submarines. Andy Karam noted in 2005,

> During exercises with Japanese diesel submarines (I believe it was during the 1988 Team Spirit exercises), *Plunger* had some problems that led to our

being beaten several times. We eventually learned how to fight against diesel boats, but by then, we probably would have been sunk. Part of the problem was the inherent quietness of diesel boats that made them very hard to detect on sonar. In addition, the Japanese crews were very disciplined—I got the impression that, if told to go to their bunks and stay there without moving, the crew would have done so indefinitely, without complaint and without breaking discipline.[53]

The Japanese tradition of strict naval discipline goes back a long way. In the 1920s, "Foreign observers noted that even when Japanese ships were in dock, sailors not on duty were kept constantly busy with calisthenics. 'We never dared to question orders, to doubt authority, to do anything but carry out all the demands of our superiors,' recalled one former seaman."[54]

The Chileans deserve to be on the list too, as their diesel submarines have successfully attacked U.S. Navy ships during exercises. In 2001, the unusually candid skipper of the nuclear submarine USS *Montpelier* (Cdr. Ron LaSilva, USN) recounted that a Chilean diesel submarine "shot him twice during successive exercise runs." As a result, LaSilva learned that "bigger and nuclear is not always better."[55] Commander LaSilva should be commended for his courage, for as we shall see later on, this kind of honesty is usually not the best policy for U.S. Navy officers.

Interestingly, that same year, a Pakistani submarine also tried to approach an American amphibious ready group operating in the Arabian Sea.[56] So many other minor naval powers have done it, as we have seen here, so why shouldn't the Pakistanis take a crack at it? Fortunately, though, this time one of the escorts, a Canadian frigate, detected the sub and escorted it from the area. This is a good thing, of course, but it still raises a question for many American civilians, namely, what exactly was a Canadian ship doing in a U.S. Navy amphibious ready group—which is now called an expeditionary strike group (ESG)? Surely, the world's largest navy can fight its own battles, yes? Well, for years, Canadian ships have been integrated with U.S. Navy ESGs and carrier strike groups (CSGs), but the rationale for this arrangement is not purely political, as some might automatically suppose, nor is it tokenism. It has much more to do with the pronounced shortage of U.S. surface combatant ships in the post–Cold War era (thanks, in no small way, to the U.S. Navy's dogmatic obsession with big-ticket supercarriers and huge nuclear submarines).

This shortage would also tend to explain why, contrary to popular opinion stateside, the U.S. Navy, at least numerically, did not play the truly dominant naval role in the Persian Gulf during Operation "Desert Storm." Morin and Gimblett have stated, "These other (non-USN) naval forces have often been overlooked or dismissed as lesser participants because, when taken individually and then compared with the American naval deployment to the region, they looked insignificant. Even the British and French task groups were small. Taken collectively, however, the other forces totaled nearly fifty ships, approximating the American effort." Indeed, "out of the total vessels dedicated to sanction enforcement, the Americans accounted for only one-third (15 out of 45), and even then the cruisers and destroyers were charged primarily with providing defence against air attack, effectively reducing their availability for other tasking." They concluded, "The relative balance of forces at sea between the U.S.N. and their allies meant that the Americans did not enjoy the same dominant position on the seas as they did on land."[57] Additionally, as we will see later, the fact is that Canadian ships are more capable in certain areas than are U.S. ships. On the bright side, though, current plans include a substantial number of new surface combatant ships for the U.S. Navy, which if fulfilled, might somewhat lessen (but probably not eliminate) the need for foreign assistance.

Retuning to our friends from Chile, in 1998, *U.S. News & World Report* noted, "In two recent exercises with Latin American navies, a Chilean sub managed to evade its U.S. counterparts and 'sink' a U.S. ship."[58] To be more specific, during RIMPAC 1996, the Chilean submarine *Simpson* was responsible for sinking the carrier USS *Independence* (this event was chronicled in the 1997 Discovery Channel TV documentary "Fleet Command"). In a 1998 article, Robert Holzer, the outreach director at the Office of Force Transformation, provided more detail: "A Chilean diesel sub penetrated the perimeter of a U.S. Navy battle group and moved among its ships for several days. U.S. forces knew the sub, participating in an exercise with the Navy, would operate in an attack mode. Yet the Pacific Fleet could not find it. The Chilean sub demonstrated that it could have targeted and fired on U.S. Navy ships at any time. In exercises over several years, the U.S. Navy's most advanced antisubmarine warfare (ASW) ships have been unable to detect the South African Navy's *Daphne* [-class diesel-electric] subs, which were built 30 years ago."[59] To wit, in a 1995 articled cited by Benedict, "Two U.S. Navy ships reportedly exercising against a South African *Daphne*-class submarine were unable to detect it even at short ranges; a U.S. observer on the submarine commented to its crewmembers, 'There is a $1B warship above you that doesn't have a clue where you are.'"[60]

In short, the U.S. Navy would have its hands full if it had to fight diesel submarines. *U.S. News & World Report* also quoted Rear Adm. W. J. Holland, USN (Ret.), who maintained if the U.S. Navy had to deal with a hostile diesel submarine today, "it would take a month to handle that problem, including two weeks of learning."[61] Strangely, though, Admiral Holland remains completely opposed to any plan that would involve the U.S. Navy acquiring its own diesel submarines! In any event, the moral of this naval story is that the American sea service really needs "a healthy dose of humility and caution in future operations."[62]

Not surprisingly, NATO and allied diesel submariners (and probably some others who are not so friendly) are extremely confident in their ability to sink American carriers. In his 1984 book *The Threat: Inside the Soviet Military Machine*, Andrew Cockburn wryly noted that European submariners on NATO exercises were far more concerned about colliding with noisy American nuclear submarines (running fast and, therefore, blind) than about being attacked by American ships.[63] Despite the vast amount of propaganda put out by the U.S. Navy, well-run diesel submarines running on batteries are quite capable of outfoxing nuclear submarines. As former Royal Navy submarine officer Ashley Bennington said in his 1999 response to an article on the *Virginia*-class submarines: "You mention that the new *Virginia* class of nuclear submarines will easily detect diesel submarines, implying that diesels are noisy. As a general rule, however, diesel submarines, which use an electric motor that runs on batteries, are quieter than nuclear-powered subs, which constantly run coolant pumps."[64] One U.S. nuclear submariner of my acquaintance had a slightly different take on this: "More specifically, nuke boats have the 60-cycle hum from an AC electrical system, the steam noise, main coolant pumps, and the turbines and reduction gears. Even when sound-mounted, these make noise a diesel boat lacks."[65] However, he disagreed with Bennington's statement that coolant pumps must be kept running at all times. "The *Ohio*-class boats can run in natural circulation at low power; the LA [*Los Angeles*] class can do so only for emergency cooling only."[66] Former nuclear submarine officer Michael DiMercurio has noted that both the *Seawolf* and *Ohio* classes can run in natural circulation, "below 35 percent power,"[67] which certainly reduces tonal output and thus makes the submarines more difficult to detect.

This applies only to low speeds. When a nuclear submarine runs at higher speeds, as many probably would in order to stop a Chinese surprise invasion of Taiwan, for example, those noisy coolant pumps would need to run, and therein lies the problem. DiMercurio says that when a nuclear submarine runs at high speed, those coolant pumps are as "loud as freight trains,"[68] which not only makes them much easier to detect and attack but makes it much more difficult for the speedy nuclear boat herself to hear possible adversaries, such as diesel submarines waiting in ambush. Compton-Hall once remarked that a nuclear submarine running at high speed is "deaf, dumb and blind,"[69] and thus quite vulnerable. The nuclear submarine's high speed advantage is indeed a double-edged sword, for it can cut both ways if not used with great discretion.

Captain Viktor Toyka of the German navy echoed Bennington's sentiments in late 2004. Toyka said that conventional submarines, especially those with air-independent propulsion (AIP), are more difficult to detect than nuclear boats.[70] Captain Li Chao-peng of the Taiwanese navy concurred that diesel submarines are more cost-effective and still quieter than any nuclear submarines. His navy has Dutch *Zwaardvis*-class diesel submarines; in 2002 he told the *Taipei Times*: "The only advantage that a nuclear submarine has over a conventionally powered one is its endurance under the sea. . . . But a diesel-powered sub like ours is much quieter than a nuclear one." He added that the Taiwanese diesel subs can definitely "compete" with nuclear boats.[71] Another former U.S. Navy nuclear submarine officer, Robert Williscroft, stated in 2002 that although the turbines used in nuclear submarines are "much quieter" than the old coal-driven turbines of days gone by, "they still make a lot of noise," and that, by comparison, diesel submarines operating on batteries are "extremely quiet." He also noted that current German and French AIP systems are "far quieter than any nuclear/steam plant. Combined with a state-of-the art sensing system and appropriate weapons, such a sub would be a formidable opponent for any nuke."[72] All these facts tend to somewhat diminish the impact of Tom Clancy's bold assertion that in the submarine world, "everyone deeply respects the Americans with their technologically and numerically superior submarine force."[73]

To be fair, though, Capt. John L. Byron, USN (Ret.), who served in both diesel and nuclear submarines, states that right now the main advantage a diesel boat has is her captain and crew, rather than the boat herself. As he told me in 2005,

> In 1960, then-Commander, now retired VADM Yogi Kaufman (my CO in USS *Cavalla* [SSK 244], where I was a Sonarman Second Class) described submarine-vs.-submarine operations as well as has ever been done: "Two blind men in a darkened room, each with baseball bats." Having an

Nuclear submarines like the USS *Texas* (SSN 775), above, often get "sunk" by diesel submarines during free-play exercise. —*M. Angebrand, U.S. Navy photo*

acoustic advantage is useful, but it's not a brick-bat certainty that the quieter boat wins. And when both are so quiet that detection range may be within 1,000 yards, matters like problem geometry, speed, tactics, luck, and the minimum arming distance of weapons play in a large way. I've routinely seen diesels take out nukes, but it has more to do with a lack of skill mixed with hubris on the part of the SSN skipper and smart diesel guys than anything inherent in the platforms (it's the wetware). . . . Diesel vs. nuke still favors the nuke, if well operated by a smart and wily skipper. But the diesel guys tend to be smarter and wilier, so it's a pretty fair fight that can go either way.[74]

Why are diesel submarine commanders generally smarter and wilier than those driving nuclear boats? Part of the answer is based on the relative ease of operation. Diesel submarines, as Scott Shuger told us, "require fewer men, and their crews don't have to undergo as much time-consuming and expensive physics and engineering training. And if a submarine is simpler to operate, crew members can concentrate more on tactics. It's long been suspected that the Navy's all-nuke sub force is very strong on running submarines but not as strong at it should be on fighting them."[75]

Although, as we have seen here, naval officers often disagree on which type of submarine tends to be quieter, the smart ones agree that, for one reason or another, the modern diesel submarine is a worthy and dangerous foe for a nuclear submarine. As Aristophanes prophetically cautioned, "The truth is forced upon us very quickly, by a foe."[76] In this instance the foe is the conventional submarine, and the truth could be rather awful for the U.S. Navy if it is ever revealed. Karam too has opined, "Now, and when I was in the Navy, I firmly felt that we would prevail in any war, from sheer numbers if nothing else. But I also felt (and continue to feel) that any war would cost us more dearly in people and ships than need be the case. Finally, my assessment of our state of readiness when I was in the Navy and Reserves was similar to the situation that faced us prior to WWI and WWII—on paper, we looked great, but I was not sure that our administrative readiness was mirrored by our actual war-fighting readiness."[77]

Today, the U.S. Navy has no diesel submarine combatants, and this means that although the diesel submarine is a very dangerous threat, the Americans must rely on smaller allies and friendly nations like Sweden, Canada, Chile, Peru, Columbia, Australia, and others to provide this vital training. This, it can be argued, is a very serious handicap for any blue-water navy, much less the world's largest. As Shuger charged in 1989, "We currently depend on the diesel subs of our allies to perform the missions diesels do best. It's foolish to rely on the British, German, and Italian navies for our security. There are crucial scenarios in which having friends with diesels won't do us much good. The Soviets maintain missile sub patrols off both our coasts, and we can't expect Britain to help us patrol them."[78] Although there are fewer Russian subs stalking the U.S. coast now than there were in 1989, they, the Germans, and the Japanese made a nasty habit of doing so at various times in the past, and future adversaries may well do the same.

Likewise, the anaclitic (for a superpower) U.S. Navy has traditionally been very weak in mine countermeasures and has often had to ask allies for those capabilities as well (because the Americans consider it unglamorous). Remember the words of Rear Adm. Allan Smith, USN, who condemned his navy's inexcusable inability to deal with primitive North Korean mines in the 1950s? When a major amphibious invasion had to be postponed because those crafty North Koreans laid mines in its path, the cocksure U.S. Navy, which had pretty much abolished its substantial mine warfare forces, was profoundly embarrassed. Admiral Smith said, "We have lost control of the seas to a nation without a Navy, using pre–World War I weapons, laid by vessels that were utilized at the time of the birth of Christ."[79] As the years went by, the embarrassment faded away, and the Navy's mine countermeasures (MCM) assets dwindled and atrophied, many of them eventually stowed away in the low-profile Naval Reserve. And there, they were ignored. Colonel Hackworth has recalled that in 1994, just prior to a visit to Korea, "I heard the Chief of Naval Operations admit that mines could be a showstopper. He was absolutely right. We don't have a competent minesweeper fleet. We still haven't appropriated money to solve this problem. But even though the admiral running the Navy had said mines could stop us, we didn't do anything about it. Instead, we bought more *Sea Wolf* [sic] subs and carriers. As I have said before, minesweepers are not sexy."[80]

When the USS *Princeton* was disabled by a mine in 1991, Rear Adm. Dan March, USN, sent a request for a ship to escort the *Princeton* back to port. The admiral specified that the escort ship had to have a helicopter and "a good anti-mine capability"; interestingly, he also said, "I'd prefer it to have a Canadian flag flying from the stern."[81] The Canadian destroyer HMCS *Athabaskan* and her Sea King helicopters quickly obliged and were dispatched to husband the American ship back to port. Now, then, why exactly did the American admiral specifically want a *Canadian* ship to do the job? According to Commodore Duncan Miller, Canadian Forces, the Americans knew that the Canadian ships "were the best prepared of any of the warships in the Gulf to counter the mine threat."[82] The Canadian ships had another advantage over their American friends, in that their Sea Kings, although not dedicated MCM platforms, were the only helicopters in the Gulf with forward-looking-infrared sensors for night missions and that truly specialized in low-level flying. Luckily, they were good enough for the job at hand, and they were the best available.

British MCM units in Desert Storm were also much better than those of the Americans. Marolda and Schneller said in 2001 that "in many ways the British flotilla was more modern and capable than the U.S. force. The five British minehunters, with Lynx and Sea King helicopters used as mine spotters, carried some of the most advanced mine countermeasures equipment then available and reflected the longstanding focus on mine warfare of the Royal Navy. British expertise in mine countermeasures was second to none."[83] Because of this proven and demonstrable expertise, British MCM forces played an integral part in neutralizing the mine threat in the Persian Gulf; initially, however, the Americans did not consult the British in planning MCM operations. After receiving requests for information, the Americans finally got around to inviting British officers to participate, and the British liaison officers were quite taken aback when they heard what the Americans had actually planned. "Commodore Craig was 'unpleasantly surprised' to learn that the Americans had developed an MCM plan to which the British had not been privy and which he considered suicidal," noted Marolda and Schneller. "He later remarked that the American concept seemed 'so ill-conceived and immature' that he 'could not believe that the U.S.N had been the architect.'"[84] One could point out that the good commodore would have been far less surprised if he had studied U.S. naval history more closely, as I hope this critique demonstrates.

During Desert Storm, which was a war about oil—let us be completely candid—mines were a major threat to shipping, and the removal of this threat was key, for the safety not only of the international naval forces in the Persian Gulf but of the tankers supplying the economies of the United States and it allies. In this vital area, the U.S. Navy's performance was nothing short of disgraceful. Adm. Stanley Arthur, USN, lamented that "everybody in the world had better minesweepers out there than I did."[85] Gen. Walter Boomer, USMC, also decried the fact that "The U.S. Navy is impotent when it comes to mines, almost, and we are very close to it on the ground side."[86] Thankfully, though, smaller allies, countries like Belgium, France, and the United Kingdom, were able to do most of the work in the North Arabian Gulf. "The allied units destroyed 1,288 mines, or virtually all the mines thought to be in Persian Gulf waters," said Marolda and Schneller,[87] but of those, just 248 were credited to the U.S. Navy, which could not have done the job by itself under any reasonable circumstances. What does this mean? It means that although the U.S. Navy was the single largest naval participant, its contribution, although obviously vital, could not be categorized as truly dominant or preeminent in all aspects. For example, after the war ended the ability of the free world to import oil safely from the Persian Gulf was largely ensured by the work of smaller allies like Belgium, whose navy actually destroyed more mines than did the Americans.

All told, said Cdr. Frank G. Coyle, USN, since the end of World War II, "14 U.S. Navy ships have been sunk or damaged by mines—more than triple the number damaged by air and missile attack."[88] Did the Navy learn from these unpleasant remedial lessons in mine warfare? As of 2004, "Our current mine warfare force consists of 14 *Avenger* (MCM-1)-class minesweepers, 12 *Osprey* (MHC-51)-class coastal minehunters, and two Squadrons of MH-53E helicopters."[89] All of these platforms are quite good, actually, but one wonders if there are enough of them to meet the needs of the world's largest and most globally involved navy. That is not a very big force, considering that America has a very long coastline and that the U.S. Navy is supposed to be the guardian of the oceans and primed for combat in the littorals. And it is unacceptable to make excuses or rationalizations, such as

"Well, our allies are supposed to do that," which, for a superpower, is really just passing the buck. As one of my colleagues said recently, "A properly balanced blue-water Navy should have sufficient MCM capabilities instead of relying on its allies or coalition partners to provide them." Besides, allies may not choose to cooperate in U.S.-led military operations if they do not think it is in their interest to do so. Indeed, Belgium decided against participating in Operation Enduring Freedom. France and Germany also abstained, and these three nations accounted for almost 45 percent of the mines destroyed by the allies during Operation Desert Storm. In fact, given that President Bush was apparently wrong about Iraq's supposed weapons of mass destruction, it might be a good assumption that many allies will be even less likely to assist the United States in future conflicts.

In contrast to the U.S. Navy's less than dominant MCM force, the much smaller Royal Navy, with a much shorter coastline to protect, has three MCM squadrons, with a total of twenty-one minehunters and minesweepers. Unlike the Americans, all British MCM units are in the regular navy, not the reserve. The Americans also have destroyers and other helicopters that offer organic mine-detection and/or sweeping capabilities, but they are not dedicated MCM platforms and therefore are probably not especially well trained in that specialty. The same goes for the new Littoral Combat Ships (LCS). The Navy recently signed contracts to build the first LCS units, and these ships can be customized with one of three mission packages, one of which is MCM.[90] These ships will be fast enough to keep up with carrier battle groups, which is commendable, but they will not be permanently dedicated to MCM (at times, they might be configured for ASW or antisurface warfare). So, in a nutshell, yes, the U.S. Navy is getting new MCM ships, but apparently only on a contingency basis, so in all likelihood they too will be far less than ideal. One of my colleagues recently said that the MCM situation is in fact much worse now than it was ten or fifteen years ago as regards the Navy's doctrine and training for neutralizing the growing mine threat in the littorals. Many believe that the U.S. Navy's recent decision to embrace organic MCM is the wrong move, as it will do nothing to rectify deficiencies in countering the mine threat and it might create a dangerous illusion that there is no need for dedicated MCM forces. The most recent Base Realignment and Closure report suggests moving surface MCM forces to San Diego and MCM helicopters to Norfolk, Virginia (both are now based in Texas). This would be another bad decision. The fact is that MCM ships and helicopters should never have been based in Texas; the U.S. Navy really needs adequate MCM capabilities on *both* coasts.

4

ASW: A Low Priority?

Our ASW capabilities can best be described as poor or weak.

—Vice Adm. John Grossenbacher, USN, 2002[1]

ASW officers and enlisted men are more often treated like the Rodney Dangerfields of the air wing. They get no respect.

—George C. Wilson, onboard the USS *John F. Kennedy*[2]

In the last section, I presented a list of U.S. Navy ships and submarines that had been "sunk" in free-play exercises by diesel submarines. As any expert will tell you, however, there are a great many variables in ASW, and to be fair, it is quite possible that, because of adverse acoustical conditions, no navy in the world could have found some of those diesel boats. And while U.S. submarines may sometimes use noise augmenters (NAUs) to simulate Soviet/Russian submarines, which until recently tended to be much noisier, there is no published evidence that this was the case in any of the exercises cited. Further, the use of NAUs is not especially common in multinational exercises (they are most commonly used in U.S. national exercises for training surface ships and P-3 crews). Several Canadian ASW senior officers I consulted (with almost seventy years of combined service) indicated that they had *never* conducted an exercise with a U.S. nuclear submarine using NAUs or operating under a handicap. One of them said of U.S. Navy nuclear submarines that, "Their job was to stay quiet to avoid us most of the time, even during exercises. Sounding like a Soviet would certainly have attracted our attention!"[3]

In the few instances in which his nuclear submarine was asked to simulate a Soviet sub during multinational exercises, Karam said that the ship "simply operated as normal (i.e., without rigging for ultra quiet)."[4] Also note well that diesel submarines often use NAUs in exercises as well—to make it easier for U.S. ships, aircraft, and submarines to find them (more on that momentarily). Furthermore, most of the diesel versus nuclear submarine scenarios described occurred in the last five years, by which time Russian submarines had made great strides in quieting.

In a 1994 article in the Russian newspaper *Krasnaya zvezda,* a Russian official made it clear that the days of American undersea supremacy were long over: "The public is told again and again that our submarines are tens of times noisier than U.S. submarines whereas their hydro acoustic equipment is 25 times inferior. Allegedly, this is the reason why they can be detected thousands of kilometers away anywhere in the ocean and will be immediately destroyed in a combat situation and therefore are not worth the expense for their construction and maintenance. I categorically disagree with such statements." Furthermore,

> The authors of these publications use for their assessment the situation which developed in the 1970s as a result of the advanced development in NATO countries of submarine detection equipment, such as the Stationary Underwater Illumination System (SUIS) [SOSUS] and flexible towed antennas as well as the appearance of the third generation *Los Angeles* class submarines. Indeed, during this period the SOSUS (incidentally, developed only because of the exceptionally favorable military-geographic situation of the NATO countries) was capable of initially detecting the location of our nuclear submarines of the first and partly the second generations in a circle measuring from ten to a hundred kilometers in radius, whereas *Los Angeles* class submarines and other maneuverable anti-submarine forces had an advantage over our submarines in the so-called 'duels.'

However,

> As for the third generation of our multipurpose nuclear submarines of the *Shark* [Akula] and *Sierra* classes (according to the NATO classification), which started to enter service in the beginning of 1980s, they are comparable in their noise level and detection distance to the U.S. nuclear submarines built at that time. With regard to our diesel submarines of

the *Varshavyanka* class, which have significantly fewer noise sources than nuclear subs, they are considered among the least noisy in the world, which earned them the name 'Killer.' All this is confirmed not only by our research institutions, Navy experts and competent foreign experts but also by the fact of collisions when U.S. submarines, while trying to establish hydro acoustic contact with our submarines, just stumble on them as if they were blind. Thus, the situation has changed in principle, the grand SOSUS has lost its effectiveness as it cannot detect our new submarines with low noise levels.[5]

As Cote has pointed out, by the mid 1980s, the Soviets had "a nuclear submarine that could elude SOSUS and frustrate efforts by tactical ASW platforms using passive sonar to establish and maintain contact with it."[6] The submarine in question, the *Victor III,* was an unpleasant surprise to the U.S. Navy when it was first encountered. A former CNO, Adm. James Watkins, described the boat "as quieter than we thought—we learned that they were hard to detect."[7] Subsequent Russian designs were even better. As Robert Moore told us, "In the early 1980s, Western naval intelligence was plunged into crisis. The Soviet *Victor III* SSNs began emerging from their shipyards in the Baltic and the Pacific and simply disappeared. Redoubled efforts to track the new class of submarine frequently ended in failure, unless the Western boat was pursuing her Russian counterpart at much shorter range. Astonishingly, the Russians had produced a submarine so quiet that they had caught up with the West. Until the *Victor III*s took to the seas, the United States thought it held a technological lead of some twenty years in terms of the reduction of submarine noise. That advantage had suddenly been lost altogether. The *Victor III*s were followed by the *Akula*-class submarines, which went to sea in the mid-1980s. Now there was no avoiding the truth. The *Akula*s were so impressive that the Russians had not just caught up with the Americans—in some key high-tech areas, they were ahead. One congressional report concluded in 1997: 'It appears that the Soviets may be ahead of us in certain technologies, such as titanium structures and control of the hydrodynamic flow around a submarine.'"[8]

Note that while the *Victor III*s were undeniably a substantial improvement over previous Soviet models, some submariners say that the capabilities of this class have been overstated. To wit, Karam said in 2005, "I remember one of our junior officers telling me that, during the 1985 spec op [special operation], he thought that the *Victor III*s were incredibly quiet and hard to track. However, during our 1986 spec op, he said that he realized they really weren't all that tough to trail, and we tracked a number of *Victor III*s successfully in 1986, 1987, 1988, and 1989. The *Akula* was a different matter—it was a genuinely quiet boat that gave us some problems the one time we went up against one."[9] Polmar said in 1997 that when the improved *Akula*-class submarine first appeared in 1990, "Admiral J. M. Boorda, the Chief of Naval Operations, told the House: 'This is the first time since we put the *Nautilus* to sea that [the Russians] have had submarines at sea quieter than ours. As you know, quieting is everything in submarine warfare.'"[10]

Was this perhaps just a case of "threat inflation," dreamed up by the admirals to extract more money from the taxpayer? I am not inclined to think so, for in 2003, DiMercurio, who usually tends to favor U.S. submarine designs over the Soviets, admitted that the Russian *Akula*-class submarine is "very capable,"[11] and earlier in his career he was candid enough to say that, at least in some ways, "The Russians were amazing and talented designers, and their submarines were the best in the world."[12] Polmar would go on to say that the Navy's claims that its new *Seawolf*-class submarine "is the quietest submarine in the world"[13] are based on highly questionable or sparse intelligence. A former CNO, Adm. Jay Johnson, told Tom Clancy in 1999 that the *Seawolf* "is awesome. The best submarine that has ever been built in the world, period. The *Seawolf* is truly, truly a magnificent submarine,"[14] but this statement must be taken with a large grain of salt. Admiral Johnson, a fighter pilot, gave this glowing endorsement after he took a ride on the USS *Seawolf* and based his statement on what he saw during his tour and from the bountiful praises the crew and other guests gave to the new boat. Of course, Admiral Johnson, and even Tom Clancy for that matter, *must* be aware that these VIP cruises are heavily scripted, methodically rehearsed, and designed for only one purpose—to "sell the product" to the brass. One can be quite certain that the crew was well coached on what to say and how to act. One can also rest assured that the submarine did not go through anything resembling a rigorous free-play combat scenario or anything else that might reveal any possible deficiencies. Indeed, the *Seawolf*-class was canceled after only three boats were delivered, but perhaps that is just as well as there were reports that these boats were not even properly tested for safety, either. In 2002, Diehl recalled that "the Navy has refused to perform shock tests on all the components of its newest type attack sub, the three-billion-dollar *Seawolf*. These supposedly required tests were designed to insure that all components would survive the stresses of most underwater explosions. The Navy apparently had diverted some of its testing funds to other uses. Such decisions continue to place those who volunteer to go in harm's way at exceptional risk."[15]

Keeping all this in mind, and unless or until verifiable evidence proves otherwise, the tired, jejune supposition that any U.S. ship that got sunk "must have been simulating a really bad old Soviet sub by running with a noise augmenter or some other handicap" should be considered knee-jerk rhetoric or wishful thinking, especially in recent years, and as such, an intellectual cul-de-sac.

Also, as we will see in this section and others, there is evidence to suggest that the U.S. Navy's ASW forces are definitely not as good as they should be, or as good as those of certain allies. This traditionally insouciant attitude toward ASW can certainly make U.S. Navy forces more vulnerable than those of other countries and, consequently, less combat effective. The Navy's standard argument on its long-term neglect of ASW is, in so many words, "We rely on our allies to help us with that"; or, "The Soviets are gone now" (which is ludicrous, considering that the U.S. Navy has been the world's largest navy for a long time now). These excuses, to state the matter simply just don't cut it.

Since the end of the Cold War and the demise of the Soviet submarine fleet, the U.S. Navy has admitted that it has not made ASW a high priority, especially in shallow water, and it shows. Perhaps the most obvious recent evidence is the demise of the Navy's carrier-based fixed-wing ASW aircraft, the S-3B Viking. In 1999, the Navy discontinued the S-3B's ASW mission, and now the aircraft are being retired, without a dedicated carrier-based fixed-wing ASW plane to replace them. Furthermore, between 1991 and 2004 the P-3 Orion force had been reduced by 50 percent, and as Goldstein and Murray[16] point out, the remaining P-3s "no longer focus on ASW as their principal mission" in most areas. For example, in 1998, Vice Adm. Richard W. Mies, USN, excoriated the laggard ASW training typical of the Navy's P-3 Orion crews. "Take the average P-3 air crew," he said. "How much time do they have on top of a friendly submarine or a potential adversary submarine just tracking them? You'll find that many of the crews have very, very little operational proficiency time. And that's true across all the elements of ASW."[17] Incredibly, Caldwell reported in 2000 that the missile-equipped Orion, a large, four-engine turbo-prop patrol aircraft, with an airframe based on a 1950s passenger plane, was actually being used to subrogate for much faster and agile jet fighter aircraft on combat air patrols).[18]

Finally, the SOSUS warning net has been "effectively mothballed," thus depriving the Navy of a much-needed early warning system. The U.S. Navy's inability to deal with quiet non-nuclear submarines was made quite evident in 2004, when Crawley wrote, "During sonar training with other navies' diesel submarines, a noisemaker or pinger is often installed to increase the sub's noise level so that U.S. warships and submarines can find the quieter vessels." The *Gotland*-class boat is "a very good submarine," added naval analyst A. D. Baker III, editor of the *Naval Institute Guide to Combat Fleets of the World* series; "Unless we enhance the [*Gotland's*] acoustic signature, we won't find it."[19]

Mind you, there is also a new initiative to improve and coordinate ASW tactics, units, training, and equipment, which includes the recent loan of one of those very same Swedish *Gotland* submarines, under the auspices of the new Fleet ASW Command (FASWC), which was stood up in April 2004. Even so, predictably, the borrowed Swedish submarine has to date proven almost impossible for the U.S. Navy to find. A Swedish periodical reported in October 2005 that after two months of exercises off the California coast, the U.S. Navy had apparently had no luck in finding the elusive Swedish boat, even though the U.S. Navy should by all means have had more intimate knowledge of the underwater geography and hydroacoustical conditions in its home waters, which are key advantages in ASW.[20] In any case, the U.S. Navy has requested Australian and Canadian help in the hunt. Of course, finding a submarine is always difficult, and perhaps the Australians and Canadians will not be able to find it either, but the request itself may be a tacit admission that others might do better. And if they do indeed have greater success in this matter (and this would not be unprecedented), once again the U.S. Navy will face undeniable embarrassment for its poor ASW skills, just as it did in years previous. Still, the jury is still out on this one, and the U.S. Navy deserves credit for trying to improve its ASW capability.

Although the temporary loan from the Royal Swedish Navy is a step in the right direction, some fear the Navy's recent reemphasis on ASW skills does not go far enough, and by that they refer to the Navy's steadfast refusal to build and develop even a small non-nuclear submarine force of its own (Admirals Zumwalt and Woodward, U.S. Navy and Royal Navy, respectively, have both recommended buying diesel boats, and even

Secretary of the Navy Lehman once said, "These submarines are extremely quiet when operated at low speeds and for this reason substantial helicopter, subsurface, and surface anti-submarine warfare defense is required").[21] They refer also to the aforementioned retirement of the S-3 Viking and the CNO's requirement that spending for the new ASW initiative must not "break the bank." One should also take note that even during the Cold War, *when there was a clear and present danger projected by a potential foe with hundreds of submarines, both nuclear and diesel-powered, the U.S. Navy was still not the most proficient navy in this specialty, even in deep water.* It remains true to this very day that other forces, such as the Canadian navy and air force, were and are arguably more committed to and more skilled in ASW (in deep or shallow water) than the U.S. Navy, despite having some old equipment like the Sea King helicopter.

A few cogent examples from history will further illustrate my point. In 1942, after several years of sitting on the sidelines of combat, America was finally at war with Germany and Japan. German U-boats prowled the Atlantic under the cloak of ocean, fog, and darkness, attacking and sinking an incredible number of Allied ships (by some accounts, they disposed of approximately twelve million gross registered tons of merchant shipping in the Battle of the Atlantic).[22] The British and Canadians had considerable experience dealing with U-boats, but the U.S. Navy initially did not want to take ASW advice from its more experienced allies. Gannon wrote that as a result of "inexperience and poor training," U.S. Navy ASW was thoroughly ineffective during the first half of 1942.[23] Not only were the Americans poorly trained, their ethnocentrism prevented them from adopting proven tactics developed by the British and Canadians. (Gannon described the American CNO, Adm. Ernest J. King, as "The Imperious Anglophobe Admiral,"[24] which is all the more interesting since, as Padfield pointed out, King's mother was born in England.[25] One might wonder what Dr. Freud would have said about this.) Germany's Admiral Karl Doenitz had a low opinion of American ASW at the start of the war: "By mid-April, the U-boats had been attacking the American east coast for three months. It appeared that their successes would continue indefinitely, and Admiral Doenitz could not resist crowing about it. 'Before the U-Boat attack on America was begun,' he wrote, 'it was suspected that American anti-submarine activity would be weak and inexperienced; this conjecture has been fully confirmed.... The crews [on anti-submarine vessels] are careless, inexperienced and little persevering in a hunt. In several cases escort vessels—coast guard ships and destroyers—having established the presence of a boat, made off instead of attacking her.... On the whole ... the boats' successes are so great, that their operation near the coats is further justified and will continue.'"[26]

Eliot Cohen and John Gooch have noted, "The Germans believed that organization and doctrine, not lack of materiel, were the roots of the American problem. The war diary contains such entries as 'enemy air patrols heavy but not dangerous because of inexperience.' '[The enemy is not] able to make allowances and adjustments according to prevailing submarine operations.' 'The American airmen see nothing, and the destroyers and patrol vessels proceed at too great a speed to intercept U-boats, and likewise having caught one they do not follow up with a tough enough depth charge attack.'"[27] Admiral Doenitz also said in his memoirs that "single destroyers, for example, sailed up and down the traffic lanes with such regularity that the U-boats were able to quickly work out the timetable being followed. They knew exactly when the destroyers would return, and the knowledge only added to their sense of security during the intervening period."[28] Add to this, one British officer noted that "there is some evidence ... that the Germans are contemptuous of A/S [antisubmarine] measures on the other side of the Atlantic and consider that the only serious restraint on U-boat operations there is imposed by the torpedo capacity of their U-boats."[29]

The Japanese too thought the U.S. Navy was badly trained in ASW. In late December 1941, a Japanese submarine prowled off the coast of northern California. It was eventually detected; however, according to her skipper, Captain Zenji Orita, "We heard a number of patrol boats, and our radiomen listened in on many plain-language uncoded message exchanges. This made it easy for us to dodge the hunters."[30] Orita later declared, "The American ASW technique at that time was very poor."[31] A German U-boat skipper also recorded, "We listened to the American radio transmissions and we heard 'We have sunk a U-boat.' We were supposed to have been sunk three times. Every time we sank a ship we were sunk again. The Americans obviously needed this as a consolation—the idea that they had done something. But it wasn't true."[32]

The British essayed to coach the recondite and insular U.S. Navy on ASW, but their efforts were rebuffed for months. "Americans must learn by their own mistakes," said Rear Adm. R. S. Edwards, USN, to a British commander, "and we have plenty of ships to spare."[33] This egregious statement betrayed a callous disregard for the safety and lives of both Americans and Allied sailors and merchant seamen. At that point the Briton told the higher-ranking American officer, "We are deeply concerned about your reluctance to cooperate and we are not prepared to sacrifice our men and ships to your incompetence and obstinacy."[34] (Otto Von Bismarck did not think much of men who were not interested in learning from the experience of others, either. In fact, he called such men "fools.") Things got even less pleasant when Admiral King had to deal with the British directly. In his 1981 memoirs, Rear Admiral Jeffry Brock, RCN (Ret.), described a noisy altercation he witnessed between Admiral King and a British admiral named Noble. King had stormed out of a meeting with Noble, looking emotionally drained from the experience. The normally calm and affable Admiral Noble then emerged from the room, also looking exhausted, and said to Brock, "I'm sorry you had to witness such a disagreeable scene. What stupid, ignorant uncouth bastards some of these people are. God preserve us from this sort of leadership if the Americans are going to be of any assistance in winning this war."[35] "Anglophobia did, no doubt, animate King in some measure," wrote Cohen and Gooch. "On a number of occasions he went out of his way to inform admirals of the Royal Navy that Britannia no longer ruled the waves and that the United States Navy was the largest and best in the world. He rejected British cooperation in the final drive on Japan (partly on logistical grounds)—a rejection later overruled, luckily for the American forces in the Pacific. It is even said that he wished to change the navy uniform in an effort to eradicate any resemblance to Royal Navy uniforms."[36] This, ironically, came from a man who, as we will see in just a moment, eventually had to accept help from the despised British and Canadians for convoy escort in U.S. territorial waters.

As noted above, the U.S. Navy's reluctance to cooperate with or listen to the British was not merely an issue in the Atlantic. British and Australian soldiers and marines had similar obstacles when dealing with the U.S. Navy in the Pacific as well. Russell Parkin offered the following:

> The reports of British officers seconded [assigned] to Australia to assist with the establishment of amphibious training, such as Lieutenant Colonel Walker of the Royal Marines, are filled with frustration at being unable to institute what they considered to be useful training. After one exercise involving the R.A.N.'s Landing Ship Infantry (LSI) HMAS *Manoora* and the U.S.N.'s APA (large amphibious transport) USS *Henry T. Allen*, Walker wrote acerbically: 'It was quite a good exercise but, all the same, there is a lot which they (the Americans) could learn from us (the British)—if only they would! . . . What worries us is the American

unwillingness to learn anything from British methods or to let the Australians and British have any say in running preparations for amphibious ops. For instance: *Manoora* was made to lower her boats empty and to go through that fatuous American boat-circling drill before the boats left to go inshore; from the beach we could hear the roar of landing craft engines five miles out to sea for one hour before the 0200 hrs [2:00 AM] landing.'[37]

During a shore bombardment mission, an Australian officer named Vickery stated that the U.S. Navy personnel on the USS *Lamson* were "unwilling to depart from set and standard ideas on procedure" and "did not seem to appreciate the Army's problems."[38]

Although Anglophobia was obviously widespread in the U.S. Navy, Admiral King was the main culprit. And, reassuringly, the British were not alone in their reaction to the man; some senior U.S. commanders also took quite a dim view of Admiral King. According to Hickam, "Even General Dwight D. Eisenhower would say of his fellow American officer: 'He is an arbitrary stubborn type, with not too much brains and a tendency toward bullying his juniors. One thing that might help win this war is to shoot King.'"[39] U.S. Secretary of War Henry Stimson did not think very highly of the U.S. Navy's senior leadership in general, either. Cohen and Gooch said that Stimson "sketched the most biting portrait of hidebound American admirals, adamant in their refusal to look at the experience of others or even common sense, men who 'frequently seemed to retire from the realm of logic into a dim religious world in which Neptune was God, Mahan, his prophet, and the United States Navy the only true Church.'"[40]

Returning to the issue of ASW, and just as an aside, Dan van der Vat has argued that unlike the Germans, the Japanese submarine force never made a concerted effort to eradicate American merchant shipping, and for that the United States should be eternally grateful. "The United States was also singularly fortunate in that the Axis seldom functioned as a military alliance in the Far East: Admiral King's troubles, had be been faced with coordinated submarine campaigns in both oceans simultaneously, hardly bear thinking about."[41] Astonishingly, "German urgings and appeals for attacks on American merchant shipping with the outstanding Japanese torpedoes (originally developed for surface vessels) persistently fell on deaf ears" in Tokyo.[42]

Like the Americans, the Canadians had extremely serious ASW deficiencies and drew derisive comments from RN officers, especially in the early years. However, the Canadians were willing to learn from the British, and by 1942 had become more efficient and aggressive at fighting the U-boats. As a result, the German U-boat commanders soon discovered that it was much easier to hunt in American waters than off the coast of Canada. While the American ASW units were acting timidly and actually blaming the British, of all people, for their U-boat problems, "by contrast, the Canadians to the north had learned from the British example and never let up on the pressure," said Gannon. "Kals in U-*139* had been so bedeviled by constant air and destroyer surveillance in Cabot Strait—surveillance as intense as any that he had experienced in the English Channel—that he moved south into U.S. waters. Bleichrodt, too, off Halifax, had been frustrated by the same relentless pressure."[43] Sarty also noted that for a time Royal Canadian Navy warships actually escorted convoys out of New York City and through U-boat–infested American waters because the much larger U.S. Navy was totally unprepared for such operations,[44] and even President Roosevelt once confided to Winston Churchill, "My Navy has been definitely slack in preparing for this submarine war off our coast.... You learned the lesson two years ago. We still have to learn it."[45] Roosevelt also quipped, "To change anything in the Na-a-vy is like punching a feather bed. You punch it with your right, and you punch it with your left until you are finally exhausted, and then you find the damn bed just as it was before you started punching."[46]

The tenacious German U-boat commanders took great liberties in the poorly protected waters off the U.S. Atlantic seaboard, insulting the hapless U.S. Navy at every possible opportunity. One such skipper, Kaleun Johann Mohr, was especially enthusiastic about hunting American ships in American home waters in March 1942: "He had not only mauled the American merchant fleet, but a recent observation had convinced him that he would have no trouble finding more targets for his remaining torpedoes,"[47] wrote Hickam. "The American freighters and tankers had begun sailing close to shore, practically steaming over the buoys that marked the dangerous shoals off the capes of North Carolina. Mohr believed there could be only one reason for this foolhardiness. The merchant masters must believe that the U-boats could not operate in such shallow water. If it had been the coast of the British Isles, Mohr knew this might be sound reasoning. *The U-boats needed depth to maneuver and hide when the Royal Navy was around. But the American Navy? Mohr only needed 30 feet of water, perhaps less, because he and the rest of the U-boat commanders were willing to attack from the surface in American waters.* This meant all they had to do was go to a buoy and wait"[48] [emphasis mine]. Admiral Doenitz concluded, "The U-boat had without question proved that it was more than a match for the defence in American waters. The same thing could not, unfortunately, be said with regards the British defensive system in the eastern Atlantic."[49]

In the first six months of 1942, as a result of the enemy's Operation Drumbeat, "an aggregate of 397 ships sunk in U.S. Navy–protected waters. And the totals do not include the many ships damaged. Overall, the numbers represent one of the greatest maritime disasters in history and the American nation's worst-ever defeat at sea."[50] In return, the U.S. Navy was able to sink only six U-boats![51] (During the same period, the British et al. were credited with thirty-two U-boat kills.) So dire was the situation that at one point Gen. George C. Marshall, USA, wrote to Admiral King to say that "'another month or so of this' would so cripple their means of transport they would be unable to bring U.S. forces to bear against the enemy."[52] Indeed, the Germans had a very good chance of disabling the entire U.S. East Coast, as Hickam tells us, if only Hitler had permitted Doenitz the required numbers of U-boats and the time to do it. Doenitz later recalled, "It is perfectly clear that 'Drumbeat' could have achieved far greater success, had it been possible to make available the twelve boats for which U-boat Command asked, instead of the six by which the operation was carried out. Good use, it is true, was made of this unique opportunity, and the success achieved have been very gratifying; we were, however, not able to develop to the full the chances offered us."[53]

If that had happened, Hickam speculated that the losses to the United States "might have proven terminal."[54] During those deadly months of 1942, "the American Atlantic coast no longer belonged to the Americans. It quite literally had become the safe hunting ground for the U-boats of Nazi Germany,"[55] says Hickam, with U-boats destroying U.S. ships "just a few miles off Norfolk, practically within sight of the American fleet."[56] Doenitz told a reporter in 1942, "Our submarines are operating close inshore along the coast of the United States of America, so that bathers and sometimes entire coastal cities are witnesses to that drama of war, whose visual climaxes are constituted by the red glorioles of blazing tankers."[57] By the end of June, Capt. Wilder D. Baker, USN, finally said something about his Navy's poor showing in the Atlantic—"The Battle of the Atlantic is being lost."[58]

After much destruction, the seemingly intransigent Americans began to listen to the British and Canadians, but only because of direct orders from the president himself,[59] who told Admiral King to establish a convoy escort system, as the British had long suggested. Miller notes that when the U.S. Navy "engaged in its first convoy battle ... the inadequacy of its training became abundantly clear. Five American destroyers were sent to reinforce the escort of SC 48, a fifty-ship convoy that had lost three ships to a wolf pack on October 15, about four hundred miles south of Iceland. By sunset the following day, the division of American destroyers arrived, and they took up station close to the convoy, a tactic that, as the Royal Navy had discovered, allowed the U-boats to launch long-range attacks with relative impunity." He continued, "Three more ships were torpedoed that night, and the inexperienced Americans panicked. They dropped depth charges indiscriminately and fired star shells, which blinded the lookouts and added to the confusion. Swinging out to avoid a Canadian corvette in the melee, the U.S. destroyer *Kearny* was silhouetted against a flaming tanker. *U-568* fired a spread of torpedoes at her."[60]

At the beginning, British ships were far better equipped and trained for ASW and convoy escort than the U.S. Navy's ships. In 1939, said the late Harvard professor Samuel Morison, only about sixty U.S. destroyers were equipped with sonar, whereas the RN had 165 such destroyers, plus fifty-four other ships.[61] It is hard to believe that this could happen in a Navy that was declared destined to be "second to none" by President Wilson back in 1916.[62] Morison also noted that in 1942 "the British were still far ahead of us in the use of asdic (as they called sonar); their sound operators on escort vessels were giving their officers more and better information than was supplied by ours. An American observer at the British Anti-Submarine School at Dunoon reported in June 1942 that lack of a specific, standard operating procedure for search and attack, such as the Royal Navy had had for some time, was the outstanding cause of our weakness in anti-submarine warfare."[63]

Interestingly, Morison (an American) referred to "asdic" as the British term for "sonar," when in fact it was the British who invented ASDIC (for Anti-Submarine Detection Investigation Committee), in 1917, then gave it years later to the Americans, who in turn gave it their own acronym, SONAR (for "sound navigation and ranging"). Brock, a Canadian who had served both in the RN and the RCN, noted that because of this change of names and the fact that American-made sets were proudly labeled "Made in USA," many Americans came to believe, incorrectly, that it was an American invention rather than a British one—whilst, I am constrained to point out, the earliest ASW sonar set was actually developed, in 1915, by the French.[64] "Our asdic gear, called today sonar, and used for locating submerged submarines, had been invented by the British Navy some twenty years before, and was still considered a secret. In fact we did not hand this invention over to the United States Navy until after they joined us after Pearl Harbor,"[65] said Brock. The Americans apparently already had similar systems, but they were obviously not as good, hence the transfer of technology from Britain. "The same thing happened with our radar, which we had originally invented and deployed, calling it RDF. It is highly probable that the modern U.S.N officer believes that these modern essentials of maritime warfare were invented in his own country."[66]

It took time, but eventually the Americans were able to hold their own with the British in ASW prowess. Indeed, the U-boat commander Reinhard Hardegen reminisced about how "surprised" he was to see that "the Americans had begun to develop defences; some ships were beginning to sail without lights, and the coast was only brightly lit close to the big seaside resorts."[67] U.S. Navy ASW skills improved dramatically (but temporarily), and that fact is frequently played down in the United States that the British and Canadians, in fact, conducted most of the ASW operations in the Atlantic. For its part, the U.S. Navy is credited with destroying 127 U-boats at sea from 1941 to 1945, and that is a very high number indeed. But it pales in comparison to the work done by the British and Canadians from 1939 to 1945.[68] The combined British/Canadian total was 491. Canada started the war with a navy of only eleven ships, five of which were minesweepers, and just 1,800 men in the regular Navy, but by the end had accounted for the destruction or capture of nearly fifty German submarines. The U.S. Navy began the war with over 337,000 personnel and more than 300 ships. Thus it is an overstatement of the highest order for America or the U.S. Navy to take sole credit for winning the Battle of the Atlantic or for defeating Germany. I say this only because many Americans have been taught that it was so. Admiral Sir Max Horne, RN, who served as Commander-in-Chief Western Approaches during World War II, was said to have been "cynical about American methods and a little resentful about the way in which the American press and other news media continued to ascribe the credit for our successes to the efforts of American forces alone."[69] Lund concludes, "The Battle of the Atlantic was fought essentially by the British and Canadian navies and air forces, although invaluable assistance was received from the United States anti-submarine aircraft and 'hunter/killer' groups during April and May 1943 when the battle was at its peak."[70] If not for the British and Canadian navies and air forces and of course British code breakers, people like Gen. George S. Patton would never have even made it to Europe in the first place.

Despite this "less than overwhelming" performance, the U.S. Navy did not seem to have a clue that the Canadians and British were far more significant players in the Battle of the Atlantic. Indeed, the U.S. Navy was hesitant to relinquish its control of convoy escorts in the northwest Atlantic under Commander Task Force 24 (CTF 24) in 1942, *even though the U.S. Navy did not contribute any ships to that sector*. When Canadian officers asked to take over command of this sector, since they were doing far more work than the Americans were, Admiral King reluctantly agreed but suggested that the American officer already in command, who had learned everything he knew about convoy escort from Canadian officers, remain as an advisor, and he suggested the Canadians were not yet ready for the job.[71] In "The Royal Canadian Navy's Quest for Autonomy in the North West Atlantic: 1941–1943," W. G. D. Lund reports that Canadian officers were incensed by this reply, annoyed by the U.S. Navy's presumptuous and baseless "know-it-all" attitude,[72] and criticized "the inefficiency of CTF 24's staff in convoy control."[73] He further declares, "It appears that Admiral King was completely out of touch with the realities of the situation in the North West Atlantic and blatantly ignorant of the scope of the RCN's participation in escort work since September 1939." A Canadian admiral then suggested that the Americans should be bluntly reminded that the RCN had been doing convoy escort for three and a half years, whereas the U.S. Navy had been in the game only for about a year and was certainly not the major contributor.[74] Eventually, the U.S. Navy relented, and the Canadians were given command of the North West Atlantic sector in April 1943.

Not surprisingly, much of what the American public was told about U.S. Navy ASW performance in the Atlantic was outright fabrication, said Regan. "If the propaganda campaign next launched by the U.S. Navy had been perpetrated by Josef Goebbels in Germany or Josef Stalin in Russia, Americans would have nodded sagely and reflected on the virtues of democracy and a free press. Instead the campaign was all-American and was used to conceal the failures of the same navy department and of its leader, Admiral King. Basically, the Navy department began issuing lies. They claimed twenty-eight U-boats had been sunk off the east coast whereas the correct figure was nil."[75] Regan summarized that "the Navy PR officers were not so easily defeated as their anti-submarine operation,"[76] in what amounted to a vast spin campaign to protect negligent senior admirals from public disgrace and possible dismissal.

As Sadkovich has said, at the end of World War II, with both Germany and Japan defeated, the U.S. Navy emerged as "the most successful navy ever—although its success clearly owed something to the British and Canadians."[77] Even Morison applauded the Royal Canadian Navy's contribution in the Battle of the Atlantic: "Too much praise cannot be given to that gallant, efficient force of our nearest neighbor."[78] The Imperial War Museum in London went even farther, one of its publications saying, "Without the Royal Canadian Navy, the Battle of the Atlantic could not have been won."[79] The American historian Ronald Spector also concludes that although few Canadian ships participated in the key North Atlantic battles of April–May 1943, "had it not been for the brave and tenacious efforts of Canada's largely amateur sailors, who at times were providing nearly 40 percent of all North Atlantic escorts, there would have been no time or opportunity to assemble the decisive components of the Allied victory of 1943."[80] Van der Vat too argues that "the worth of the Canadian contribution to the campaign has generally been rather more seriously understated"[81] and that for a country with such a small population, "its contribution was astonishing."[82] Canada was the first country to sink three U-boats in a single day,[83] for example, and it also "developed a successful night-attack training system which was borrowed by the two larger navies."[84] British admirals frequently criticized the RCN, but even one of their number, Admiral Sir Percy Noble, eventually conceded, "The Canadian Navy solved the problem of the Atlantic convoys."[85] Commander. Tony German, RCN (Ret.), wrote in 1990 that Canada also controlled "the submarine tracking and all the western Atlantic convoy routing for the first six months the U.S. was in the war—a fact yet to be acknowledged by British and American historians."[86] The little guys here clearly made a big difference, but as I do my best to be intellectually honest, one should also read Marc Milner's more critical account of the RCN's performance.[87]

As we all know, the United States supplied ships to Britain and Canada during the war, and much has been made of this in the United States. There have been many statements that U.S. ships were better equipped than those of other countries; that was true in some cases, but Brock notes that even some of the newly constructed American-made ships were, in point of fact, rather poorly designed. His ship, HMS *Bazely*, was built in the United States for the Royal Navy, and he recalled of its construction, "The United States Navy yards abounded with technical experts of every description who had never been to sea. They did not understand my reluctance to embark upon a long ocean voyage without a magnetic compass of some sort."[88] The U.S. Navy insisted that only a gyrocompass was needed, but Brock's small *Captain*-class frigate had been equipped with a gyroscope intended for a battleship or an aircraft carrier. "It simply couldn't take the punishment required of any gyrocompass in a small vessel encountering heavy seas."[89]

He also complained about the American-designed ship's ridiculously unwieldy communications equipment. The *Bazely* had 102 telephones and an automatic switchboard, whereas a similar British ship would have been able to manage with just six. "These elaborate internal communication arrangements filled me with dismay because I had no idea how we were going to keep such a plethora of flimsy instruments fully operational under seagoing conditions," he reported.[90] "I asked one of the navy yard electrical engineers what we would do, for instance, with our bridge telephones in a heavy sea when our open bridge was being washed down with salt water. His reply was simple, and I guess, logical: 'If the bastard busts, throw it overboard and plug in a new one. There'll be plenty in your central stores. We mass produce them, you know, and they only cost twenty-three cents apiece.' Such startling revelations as these in all departments of the ship gave us an entirely fresh outlook on how Americans get things done."[91]

Brock's disdain for U.S. Navy communications equipment continued even after the war, when he did his best to sideline Canadian attempts to adapt U.S. Navy systems. As he said, "As far as I was concerned, we already had a better communications system than anyone else in the world except, possibly, the Royal Navy."[92] It appears that some U.S. Navy officers might agree with Admiral Brock's criticism. Capt. Peter Huchthausen, USN (Ret.), served in a *Forrest Sherman*–class destroyer in the 1960s, and he described the ship's communications system as "a nightmare."[93] Even though this class of ship was assigned to ASW duties, he said it was poorly equipped for the task and was not a decent gunnery platform either. He also quipped that this class had "a highly dangerous twelve-hundred-pound-per-square-inch steam plant, which killed more men in accidents that any destroyer propulsion system before then."[94] Other U.S. escort ships of the 1950s and 1960s were even worse, said Michael Isenberg: "The *Dealeys* and *Bronsteins* were each built with single screws, and the *Claud Jones* were diesel-powered, austere turkeys that could not come close to adequately performing their escort function. These escorts were far too slow to tackle the fifteen-knot Type XXIs or XXVIs [diesel submarines]. Most of the escorts could not hustle much over twenty knots; none of them could boast the required ten-knot speed advantage over the diesel-electric boats."[95] As a comparison, it should be noted that the U.S. Navy's Operational Evaluation Command evaluated a Canadian escort, HMCS *St. Laurent*, in the late 1950s and concluded that she was "the best of her type ever built."[96] Indeed, U.S. ships are not always as good as the American public has been led to believe.

It is also ironic to note that the U.S. Navy also borrowed or bought warships from Canada and the United Kingdom during World War II (including an aircraft carrier, HMS *Victorious*). This last comment is a minor, perhaps trivial point, of course, but it, along with the U-boat hunting statistics mentioned above and the reality that Canadian and British ships had to escort Allied shipping through American waters, surprises many who espouse the traditional "If it weren't for us, you'd all be speaking German" polemics so often recited in certain lay circles. Actually, when the British deployed some two dozen ASW trawlers to the U.S. east coast in 1942, the British viewed it as a "rescue mission." Interestingly, when one of these British ships, the *St. Loman*, came to New York, her captain was very disturbed by the U.S. Navy's and Coast Guard's lack of vigilance. "When we were coming into New York we saw a destroyer, so we flashed her with our Aldis lamp. And she took no notice of us at all. We went in a bit closer. There were some small patrol boats and we flashed them; no one replied. We made our way up through the passage and into New York Bay, and no one seemed to take any notice of this little British ship flying the white ensign." Later, when the ship departed New York, once again she encountered a patrol boat, which also did not reply to her messages. The British commander then decided to shake things up a little by firing a starshell, "which made rather a loud bang. And everybody woke up on that boat and they headed full blast for the Jersey shore."[97] The British were not impressed, and ironically, it was about this time that Admiral King was particularly full of bluster about how much "better" his navy was than the RN.

That was World War II, but some things never change, or they change only temporarily. In fact, "By 1958, the CNO, Adm. Arleigh Burke, wanted 'to know why the Navy's ASW effort, despite all the high tech, was so weak and ineffective.'"[98] In 1959, during the U.S. Navy's first test of a carrier task force's ability to deal with nuclear submarines, the submarine USS *Skipjack* was chosen to play the enemy. According to Lieutenant Commander Stuart Soward, RCN (Ret.), the *Skipjack* was "unrestricted in movements," and she proved to be too much for the U.S. Navy surface ships and aircraft during the exercise. However, a single experimental Tracker aircraft of the RCN, with its new Canadian-designed ASWTNS (Anti-Submarine Warfare Tactical Navigation System), was also involved; it was the only aircraft that could detect and maintain contact with the *Skipjack*. Sadly, the Canadian tracker was not carrying any exercise weapons and could not attack; however, if it had, Soward said, the submarine would have been finished. The U.S. Navy apparently agreed and was so impressed with the performance of the Canadian plane and its systems that it almost immediately placed orders for the ASWTNS system.[99] Because of things like this, perhaps, Michael Isenberg says, the Canadians are known for being a "superb ASW force."[100]

Most Americans do not know it, but the U.S. Navy found itself once again dependent on the Royal Canadian Navy for essential ASW forces during the 1962 Cuban Missile Crisis. When Kennedy decided to quarantine Cuba, the U.S. Navy, still far superior to the Soviet Navy, at least on paper, had to seek assistance, because it did not have enough ASW escorts to do the job (a familiar story). According to Commander Peter Haydon, RCN (Ret.), two American admirals came to Canada to request assistance in dealing with Soviet submarines; the Canadians obliged by deploying much of their Navy, RCAF ASW aircraft, and the two British submarines under their control to sweep the North Atlantic for Soviet submarines. According to David Robinson, the U.S. Navy established a six-hundred-mile submarine barrier south of the Grand Banks; "It was a huge undertaking, and with American naval forces stretched to the limit with the Cuban blockade, major Canadian participation was essential to its success."[101] Later, the Canadian ships were asked to move farther south, and just as they had in the early days of 1942, they patrolled the waters approaching New York Harbor. As the historian Tony German has articulated, "The RCN took over a very substantial segment of what would have been a U.S. responsibility and certainly allowed at least one [antisubmarine] task group to move down further south."[102] During these crucial operations in the treacherous waters of the North Atlantic, the deployed Canadian airborne ASW units were superior to those of the U.S. Navy, both qualitatively and quantitatively (twenty-four RCAF CP-107 Arguses versus just seventeen U.S. Navy P-2V Neptunes, which had older equipment, shorter range, and less comprehensive training in ASW). In fact, "Commodore J. C. O'Brien, the Canadian naval attaché in Washington, knew that heavy American commitments in the Mediterranean and Pacific meant Canada 'had more ready forces in the ASW business in the Atlantic than the Yanks did.'"[103]

ASW: A LOW PRIORITY?

All told, between October 23 and November 15, 1962, the RCN made 136 contacts with Soviet submarines and forced a number of them to surface, without the help of SOSUS.[104] Haydon has noted that "Canadian maritime forces became responsible for most of the anti-submarine warfare (ASW) operations in an area that stretched from Greenland to New York harbour."[105] In gratitude, the U.S. Chief of Naval Operations (Adm. George Anderson, USN) and his wife came to Canada to have a private dinner at the home of the Chief of Naval Staff, just to say "thank you" for the help.[106] It's great to have allies, and the RCN also depends greatly on the U.S. Navy (obviously), but one must wonder why the world's greatest navy was not able to fight its own battles *in its home waters*, no less, not once, but twice since the beginning of World War II.

It would be unfair of me not to mention that the U.S. Navy also enjoyed great success against Soviet submarines during the Cuban Missile Crisis, with various U.S. destroyers and aircraft chasing down Soviet ballistic missile submarines and forcing them to surface. These feats are well documented in Captain Huchthausen's 2002 book *October Fury*, and no doubt many Americans are aware of this. What they may not be aware of is that the U.S. Navy made at that time many of the same clumsy mistakes that it made in World War II. As I mentioned earlier, Japanese submariners were amused that U.S. Navy ASW forces did not bother to encode their messages in 1941, and Russian submariners made exactly the same comment almost twenty-one years later. Aboard the ballistic missile submarine *B-36*, "[Lieutenant] Zhukov also noticed that U.S. pilots were extremely careless with their communications, and often in the heat of prosecuting a possible submarine contact they departed from the lightly coded terminology and returned to uncovered, clear UHF and VHF voice communications, which Zhukov and his radio intercept operators copied without difficulty. Zhukov found that the aircraft carriers *Essex* and *Randolph* were the worst offenders of radio discipline, and resorted often to clear communications. . . . It was amusing for the Russians to hear the pilots using nicknames and their abbreviated call signs to conceal their identity. Any half-witted intercept officer could easily pick out the various pilots by listening to their talk."[107] Huchthausen also noted, "By carefully triangulating their radio transmissions, Zhukov, [Captain] Dubivko, and the other watch officers were able to keep tabs on the destroyers and remain on the edge of the hunter's search patterns."[108]

The American ships, ill equipped for ASW and amateurish in their tactics, won the day in spite of themselves, but mostly because the hydroacoustical conditions in their sector were quite favorable to them, leaving the Soviet submarines with no thermal layers under which to hide. The Soviet submarines were also under orders to run at high speed for much of their cruise across the Atlantic, which obviously compromised their ability to remain undetected.[109]

Thus far, we have reviewed the U.S. Navy's relatively poor and sloppy ASW tactics and equipment in World War II, when they relied on the British and Canadians to a great extent, and during the Cuban Missile Crisis, but the embarrassment does not end there. If we move ahead now to the early 1980s, we see that Canada's cheese-paring, antimilitary federal leadership had allowed the Canadian navy almost to rust out, yet due to its intensive training and emphasis on ASW excellence, it was *still* better at hunting submarines than the U.S. Navy. In 1983, a retired British naval officer and senior fellow at the Brookings Institution, Mike McGwire, told a Halifax newspaper that "ship-for-ship," the Canadian Navy's elderly ASW destroyers were still "better equipped, maintained, and trained" and "infinitely better" at ASW than American surface ships.[110] At the same time, the new Canadian CP-140 Aurora aircraft was arguably far superior to its elderly cousin, the American P-3 Orion, and the *Oberon*-class submarines were arguably much better listening platforms than the U.S. Navy nuclear submarines of the time. During his service on the elderly nuclear submarine USS *Plunger* during the late 1980s, Karam recalls, "We were almost never detected during games with our own Navy, and then only when we approached on an agreed-upon bearing at a given time and usually cavitating or going active on sonar. I took many photos of our surface ships at close range at a time when they were unaware of our presence. . . . *Plunger* made successful attacks against U.S. carriers, cruisers, destroyers, frigates and a battleship during my time on-board."[111]

Regrettably, not much has improved in the U.S. Navy since. For example, in the late 1990s Capt. D. Michael Abrashoff, USN (Ret.), then a commander, participated in a three-ship ASW exercise against a U.S. nuclear submarine in the Pacific. (The surface ships were the USS *Benfold, Gary,* and *Harry W. Hill.*) Abrashoff developed a very thorough and creative plan to deal with the submarine, but his plans were rejected by his superiors "in favor of a last-minute plan based on the same tactics the Navy had been using since World War II. . . . As a result, the submarine sank all three of us—without its crew breaking a sweat."[112]

Despite all the practice they had had with Soviet submarines during the Cold War, U.S. submarines, ships, and aircraft today are still often not as well prepared for ASW as Canadian units. Today, Canada's incoming *Victoria*-class diesel submarines (formerly the British *Upholder* class), *Halifax*-class frigates (also known as the Canadian Patrol Frigate, or CPF), both of which are or will be equipped with the AN/SQR-501 CANTASS towed sonar array system, are world-class ASW assets. Deployed in the mid-1990s, CANTASS wedded the existing U.S.-designed SQR-19 towed array with a "breakthrough technology" Canadian processor (the AN/UYS-501); at that time both American and Australian naval officers called it the best in the world, as was demonstrated in many naval exercises.[113] Both of those countries subsequently bought variants of this processor from Canada. In addition, the modernized Tribal-class destroyer, which was "the first all gas-turbine propelled ship in the NATO and the first frigate to fly two all-weather helicopters,"[114] and the updated CP-140 aircraft are or will be in many ways better equipped, better designed, more suitable, and better trained for ASW than their current American equivalents, although the U.S. Navy's P-3 Orions will soon be replaced, if all goes well.

If one compares contemporary U.S. nuclear submarines to the small Canadian diesel submarine fleet today, again one finds reasons to believe that the Canadians have better undersea ASW platforms. For example, contrary to the misleading agitprop now in circulation, some diesel submarines have excellent weapons systems, comparable to those found in the much more expensive nuclear submarines. If we put aside the teething problems currently affecting this class, Compton-Hall has said, the Canadian *Victoria*-class submarine has "an exceptionally good weapon system, equivalent to an SSN with Ferranti-Gresham-Lion DCC fire control. . . . [T]he submarine is extremely quiet."[115] David Miller, formerly of Jane's Information group, postulated in 2002 that the Canadian submarines "are the most sophisticated and capable diesel-electric submarines ever built."[116] Said Commander Jonathan Powis, RN, who commanded one of the *Victoria*-class boats while they were in British service:

> The greatest strength of the class is its small acoustic signature. Benefiting from 35 years' money and effort expended in quieting nuclear-powered submarines, they are extraordinarily quiet. On main motor they were shown repeatedly to be all but undetectable by passive sonar. Even when snorkeling, they had a signature comparable to a modern SSN. . . . They presented a difficult target to active sonar as well because they were small and fully acoustically tiled, and much of their superstructure was made from composites. Moreover, because of their size, adversaries could not

easily exploit magnetic anomaly detection and other nonacoustic signatures.[117]

A small quiet diesel submarine can be exceptionally good at detecting an enemy, while not being detected herself. The same cannot be said of U.S. Navy nuclear submarines, which are considerably larger and, depending on what submariner you choose to believe, perhaps somewhat noisier. During RIMPAC 2004, for example, a Canadian CP-140 detected and tracked the redoubtable nuclear submarine USS *Charlotte* using sonobuoys and its magnetic anomaly detector. The ostensibly mighty *Charlotte* was depicted simply as "a huge metal object disrupting the earth's magnetic field." The initial stage of the hunt was scripted (the Canadians knew that the *Charlotte* was at or near the surface in a specific area), but after submerging the *Charlotte* did her best to evade the plane, even trying to leave the designated exercise area. Nonetheless, the Canadian plane was able to maintain contact and track the submarine. As one pleased Canadian officer reflected, "This was good training. . . . [W]e had him early and we held him at an extended distance." The *Charlotte* tried to shake the patrol plane, but she did not succeed.[118] As the submariners say, "Aircraft, mark on top!"[119]

Fortunately, these ASW failures and shortcomings are finally and slowly becoming public knowledge in the United States, for as the Congressional Budget Office revealed in 2001,

> Some analysts argue that the Navy is not very good at locating diesel-electric submarines, especially in noisy, shallower waters near coastal areas. Exercises with allied navies that use diesel-electric submarines confirm that problem. U.S. antisubmarine units reportedly have had trouble detecting and countering diesel-electric submarines of South American countries. Israeli diesel-electric submarines, which until recently were relatively old, are said to always 'sink' some of the large and powerful warships of the U.S. Sixth Fleet in exercises. . . . Thus, if a real opponent had even one such submarine with a competent commanding officer and crew, it could dramatically limit the freedom of action of U.S. naval forces in future conflicts.[120]

For more on the relative deficiencies of U.S. Navy ASW, please see chapter 11, "Lack of Training, Overrated Technology, Bad Policies, and Technocratic Leadership."

5

A Lucky Break at Midway and the Big-Carrier Navy

I'd rather be lucky than good.

—Vernon "Lefty" Gomez[1]

SURELY IT IS FAIR TO SAY that foreign navies openly and unashamedly flaunt it when one of their submarines "sinks" an American carrier in an exercise. They have no problem letting the news media know about their triumphs. With a few courageous and candid exceptions, such as the people quoted in this book, American nuclear submariners generally do not *publicly* reveal their own accomplishments against U.S. Navy aircraft carriers. If they do, they do it anonymously, usually after they leave the service, or they provide only the sketchiest of details. Why is this so? Former U.S. Navy officer Jerry Burns gave a pretty straightforward answer in 2000—because "anyone who says something is wrong gets thrown out of the Navy."[2] Also, as Professor Thomas Etzhold has pointed out, the U.S. Navy does not want anyone to know that its carriers have been sunk (or even seriously damaged) in exercises. Ergo, officers are strongly encouraged to keep quiet about such incidents. Obviously, these gag orders only apply to U.S. Navy personnel, not to foreign crews. The author of the 1987 book *War Games,* Thomas B. Allen, described this naval censorship during an interview with the American NPR network in 2003: "The Navy had a kind of unwritten rule: You can't sink an aircraft carrier in a war game. And if you talked to any submariner who had been in either an exercise or a war game, you get a whole story about how many times they really sank aircraft carriers."[3]

In 2000, Gutmann observed, "People on active duty do not tell reporters the truth if the truth is something they know their COs will not want them to say. Many, many service people have ruined or lost their careers testing this rule."[4] The Navy's public affairs officers (PAOs) closely monitor interactions between journalists and Navy personnel to ensure that no one complains or says anything that does not tow the company line. In Gutmann's experience:

> The PAO's very presence, his dogged insistence on gluing himself to the reporter's side, puts a wall between the reporter and the world he is trying to understand. When one does manage to corner an actual Sailor (with the PAO a couple of feet away, trying to appear as if he really is just suddenly concerned with the condition of his finger-nails or the patch of linoleum he's found himself standing on), the sailor will stand rigidly at attention (while the PAO is watching the sailor out of the corner of his eye), and then proceed to spout a lot of boilerplate that the reporter might as well have copied off the official DOD-sponsored Navy Web page. Going 'off script' in today's military is too often a career killer, and nobody's ready to take the risk of saying what they really think unless they've signed their resignation papers.[5]

Hard to believe, but to me this does not sound very much different (or better) than life in the old Soviet Navy, with its political officers and GRU agents aboard ship, watching and listening for any sign of free, unstructured thought. Thankfully, in the U.S. Navy, this PAO surveillance only pops up when journalists are aboard ship.

In this world in which great effort is made to conceal the truth, it is not surprising that a carrier cannot be "sunk," *even if it really did happen*, such as in 1964, when a Vietcong swimmer team sank the old escort carrier USS *Card* in the shallow water of the Saigon River. According to Dunnigan and Nofi's 1999 book *Dirty Little Secrets of the Vietnam War*, she did sink, and everyone knows it, but at the time the Navy could not bring itself to admit officially that one of its carriers, even a small one that had been converted simply to haul helicopters and planes (not operate them), could possibly be sunk so easily.[6] The headline "U.S. Carrier Sunk in Vietnam" would not look good in the U.S. newspapers, but the North Vietnamese felt otherwise and issued a postage stamp that proclaimed, "Aircraft carrier of America Sunk in Harbor of Saigon."[7] Although the ship was by no means a supercarrier, the sinking of the *Card* was especially hurtful, since she had won a

Presidential Unit Citation in World War II; instead of saying she had been sunk, the Navy said that she had only been "damaged" and quickly "repaired," rather than sunk and refloated.[8] In other words, the gist of all this is that the truth is suppressed for "the good of the service." We can therefore deduce that the good of the service is the paramount concern in the U.S. Navy, not the good of the country, and not the good of the taxpayers who bankroll these expensive platforms.

The U.S. Navy's aircraft carriers have plenty of supporters and wagtails, of course, including many politicians who cash in politically on the jobs that naval contracts provide to their constituents. One of their most common defenses is to invoke a sophism and imply that since no big American carrier has been sunk since World War II, America's big carriers cannot be sunk. This is tantamount to saying "My residence is in one of the most dangerous areas of Washington, D.C., and in sixty years it has never been burglarized. The only possible explanation is that it must be burglar proof." The real reason that no big American carrier has been sunk in the past sixty years could simply be that no one in the area had the motivation, necessity, or opportunity to try. Every time I hear this specious reasoning or some variation of it, I quote the late Newton D. Baker, secretary of war in the 1920s. Baker scoffed at Billy Mitchell's claim that air-delivered bombs could destroy the "unsinkable" and "invulnerable" battleship. When Mitchell suggested doing a bombing experiment with aircraft and a stationary German battleship, Baker said, "That idea is so damned nonsensical and impossible that I'm willing to stand on the bridge of a battleship while that nitwit tries to hit it from the air."[9] Had he done so, Secretary Baker might well have been killed, because as we all know now, Mitchell was right, the battleship went down, and Japanese aircraft later did even better, sinking two British battleships that were fully manned, equipped, and under way—but once again, I digress.

John Lehman was quite right when he said that no big *American* carriers were sunk during the war, but it is a fallacy to assume that this is because of some special quality of American aircraft carriers (which, by the way, were more vulnerable to kamikaze attacks than British carriers, because the American ships did not have armored decks).[10] If the Japanese had succeeded at Midway or Guadalcanal—and the U.S. Navy should thank its lucky stars that they did not—we might not even be discussing the American supercarrier today. Nor is it necessarily true that U.S. carrier task forces have been a successful deterrent either, because, as the Malayans aptly say, "Don't think there are no crocodiles because the water is calm."[11] Those carriers did not deter the North Koreans, or the North Vietnamese either, for that matter.

As former naval intelligence officer Scott Shuger reminds us, the world's largest carrier in World War II, the 71,890-ton Imperial Japanese Navy Ship *Shinano*, was sunk in 1944 "by four torpedo hits from a single American submarine"[12] (the USS *Archerfish*). According to the sophic Capt. Joseph Enright, USN, the skipper of the *Archerfish*, the *Shinano* was heavily armored and well protected from torpedoes: "The weight of the steel installed for defensive purposes totaled 17,700 tons—about one-quarter of *Shinano*'s displacement and equal to the tonnage of many light cruisers. . . . As for the watertight integrity of the ship, there should have been little cause for concern. Since 1935 Japanese warships were tested first by filling the underwater compartments with water, then, after the equipment was installed, by conducting air tests. The compartments of *Shinano*, which had been tested hydrostatically, were structurally sound, and watertight doors had been installed."[13] And even though the *Shinano* was unable to run at full speed because some of her boilers were unserviceable, she could still make more than twenty knots, faster than any diesel submarine. Like the Americans do today, the Japanese considered their first and only "supercarrier" to be virtually unsinkable, and yet four torpedoes violently disproved that claim on November 29, 1944. (If four U.S. torpedoes could do this, imagine what horrors the Japanese could have inflicted at Midway, with the best torpedoes in the world, if only they had employed their submarines as effectively as the Germans did.) This achievement lends credence to the statement of French novelist Honoré de Balzac that "power is not revealed by striking hard or often, but by striking true."[14]

Yes, the *Shinano* was not fully tested and cleared for combat duty, but the U.S. Navy has most definitely sent carriers to sea when they were less than fully ready for combat too. Williscroft attested that the USS *Independence* was far from shipshape during her deployment in early 1998 (the conditions aboard were described as "atrocious," with "critical maintenance being neglected . . . decks were waxed, but the crew was incapable of handling a real emergency.") Such large but poorly maintained ships would be relatively easy to destroy, as was the *Shinano*.[15]

Four years later, the carriers USS *John F Kennedy* and *Kitty Hawk* failed a major readiness inspection and a light-off assessment (respectively), with the *Kennedy*'s propulsion system declared "unsafe for operation."[16] In late 2001, the *Kennedy* was a deeply troubled ship that had failed a visit by the Board of Inspection and Survey (INSURV) just a few months before deployment: "Three of the ship's four aircraft elevators, used to bring aircraft from the hangar deck to the flight deck, were inoperable; two of the four catapults that launch aircraft were in bad shape, and the flight deck's firefighting equipment was 'seriously degraded.'"[17] Even worse, one of the *Kennedy*'s men said that the dilapidated ship, which required extensive emergency repairs to make ready, was not the worst ship he'd seen in his nineteen years of U.S. naval service. After the repairs were made and the *Kennedy* deployed in 2002, some Navy wives were still not sure that their husbands would be safe on such a run-down ship. "Will the *Kennedy* come home after it leaves?" wrote Karen Moore, whose husband was attached to the ship. "I worry that this ship is destined for disaster."[18] Such a disaster, arguably, is more likely if the U.S. Navy continues to deploy poorly maintained, undertrained, and undermanned aircraft carriers. The Japanese did this sixty years ago with the *Shinano*, and it could just as easily happen to U.S. ships today. (It was announced in September 2006 that the *Kennedy* was to be decommissioned). Even worse, at least the Japanese did not have to contend with attacks by nuclear-tipped torpedoes or cruise missiles, a fate that could now befall the U.S. Navy at any time.

While on the subject, U.S. Navy battleships have also been deployed in poor materiel condition, and during the U.S. Navy's palmy, halcyon days under Ronald Reagan, to boot. According to military historian Geoffrey Regan, the battleship USS *Iowa*, launched in 1942, was by then in many ways, although modernized and reactivated in 1984, an ancient vessel that was "basically unreliable."[19] By the late 1980s, as Regan put it, "The *Iowa* was not in good shape. The new captain found a loose hatch in one of the turrets that had been leaking hydraulic fluid for two years. The crew in the turret used twenty-five watt light bulbs for fear of blowing fuses if they used fifty watt bulbs. In the gun-loading areas, bags of explosive propellant were torn and were leaking black powder. Nor was the crew up to scratch, quantitatively or qualitatively. The ship was short of good petty officers, had an annual turnover of crew of forty percent, and in one turret was short of thirty-seven of the 118 men who usually served there."[20] Not only was the ship short of men, many of the men she did have were "dopers, marginal personnel," said the *Iowa's* skipper, Capt. Fred Moosally.[21] The *Iowa's* deficiencies were obvious to many of her officers, including Lt. Cdr. Dennis Flynn, USN, the director of the ship's strike warfare center. Flynn predicted in 1988 that the *Iowa* would be "sunk" in free-play exercises, and he was proven right. As former naval officer William C. Thompson II (no relation to this author) recorded in the fall of 1988, "The

The mighty battleship USS *Iowa* (BB 61) was "sunk" at least twice by Dutch, British, Canadian, and West German forces in 1988 and 1989. —*USNI Archives*

Iowa engaged NATO forces and was 'sunk' by a Dutch frigate hiding lurking behind a civilian oil tanker."[22] A few months later, in the Caribbean, the *Iowa* was again "trounced by the British, Canadian, and West German forces."[23]

In April 1989, as the story goes, the novice crewmen in the *Iowa*'s number-two turret were put on the spot during an exercise because they had to fire the guns with little or no experience with the equipment or procedures; to complicate matters even more, they had never worked together before. The resulting explosion killed forty-seven men. As journalist Peter Cary concluded, "If ever there was a ship seemed fated for catastrophe, it was the USS *Iowa*."[24] A poorly maintained ship, like the aircraft carriers aforementioned, was sent to sea too soon, and it was theoretically destroyed in two exercises, and then partially destroyed in real life. A very sad story, but a strong reminder for the U.S. Navy that even the most impressive ships can be destroyed easily, especially if they are not well maintained and their crews not well trained.

No study of this kind would be complete without reference to the USS *Pueblo* debacle of 1968, in which a poorly trained, badly maintained, poorly equipped, and overmanned U.S. spy ship was sent on a badly planned mission to North Korea and quickly captured, without a fight. The crew was incarcerated for almost a year. The United States did not retaliate, and North Korea retains the *Pueblo* as a war trophy to this day.[25] Perhaps this incident can explain why the North Koreans appear to have absolutely no fear of the U.S. Navy.

The pro-carrier argument I mentioned above loses even more strength when we consider how easily the U.S. Navy might have lost the Battle of Midway in 1942. In his brilliant work "Our Midway Disaster: Japan Springs a Trap, June 4, 1942," Professor Theodore F. Cook theorizes that had the Japanese been just a little bit more diligent and skeptical about the phony radio reports about Midway's water problems, there is a very high probability that they would have won the ensuing battle. "Given the deadly suddenness of carrier warfare," he notes, "how easily might it have been the U.S. Navy mourning the loss of three carriers . . . in exchange for, perhaps, one or two Japanese flattops on June 4, 1942?"[26] Furthermore, he recommends that his readers ponder a rather unpleasant theoretical possibility: "What would have happened if the Japanese had won at Midway? With only one carrier left in the Pacific, how could we have resisted their advance?"[27] One should never forget that the American victory at Midway was far from certain and has been often been called a "miracle." Heavily outnumbered and, much more importantly, thoroughly outclassed by pilots with substantially more flight experience and presumably much higher morale, the Americans prevailed, but this was largely due to the gullibility of a few Japanese naval personnel.

The Japanese were trained to much higher standards than were the Americans, especially in night fighting, and were better equipped in many categories, especially, said Morison, in "pyrotechnics and optics. Their starshells and parachute flares were brighter and more dependable than ours; their binoculars were so much better, especially for night work, as to be eagerly sought after by American officers and bluejackets. Their naval officers were excellent navigators."[28] Also, "The Japanese Navy conducted its battle training in remote waters where it could not be observed, and where they would be hardened by exposure to the elements. That this rigorous and realistic training under combat conditions paid off, was all too evident in the first months of the war. . . . In contrast, the United States Navy normally carried out peacetime maneuvers and exercises in southern waters or where fine weather prevailed. Extra precautions had to be taken to avoid casualties and consequent unwelcome publicity."[29] Morison also spoke of Japanese superiority in torpedo training: "Moreover, the Japanese Navy fired torpedoes freely in practice and at maneuvers, thus improving them constantly; while the United States Navy had to economize when testing warheads and exploders, and never found out what was the matter with its torpedoes until the war had been going on many months."[30] O'Connell notes:

> [A]t the time of Pearl Harbor, Japanese naval aviators were quite probably the best in the world. Hardened by a Spartan and comprehensive training program, by 1941 pilots averaged in excess of 300 hours of flight time before joining the fleet, far more than their American counterparts. Moreover, a good many had combat experience in China. In addition, the aircraft they flew were excellent. Of particular note were the torpedo planes—the long range Type 95 and the tactical 'Kate,' both being superior to their U.S. competitors. But most remarkable was the Zero, in 1940 the world's most advanced fighter aircraft.[31]

Orita, interestingly and provocatively, has declared that Midway could have been salvaged had only the Japanese properly deployed its submarines to locate and attack the American carriers: "Had our submarines been used properly and effectively, the history of the Pacific War might have been written quite differently."[32] And as Captain Mitsuo Fuchida, IJN, observed, the losses at Midway could have been quickly avenged and the vestigial components of the U.S. Pacific fleet erased if only Admiral Yamamoto had had the nerve to order his additional carriers to Midway to continue the fight instead of sending them to the Aleutians.[33] For the Japanese, a battle that should have been a cakewalk was lost because, by their own admission, it was they who made "all the errors in this action," despite overwhelming tactical superiority in most areas. This might lead some to conclude that the U.S. Navy did not so much "win" at Midway as receive the benefit of the Japanese simply botching what should have been a certain victory.

American history books record the outcome of Midway as a turning point for the Americans, but nevertheless this battle also demonstrated that despite the inadequacy and indecision of some of their leaders, the Japanese were the better warriors on a man-for-man, plane-for-plane basis, even though U.S. and British officers had initially dismissed Japanese pilots as "inevitably near-sighted" and lacking "initiative."[34] Even though they had the element of surprise working for them, the first seven U.S. air attacks against the Japanese fleet were utter failures. Walter Lord has said that the Americans took "crushing losses—15 out of 15 in one torpedo squadron . . . 21 out of 27 in a group of fighters . . . many, many more. They had no right to win."[35] Their air combat tactics were amateurish and clumsy: "At 9:36 the *Akagi* ordered cease fire; the fighters were bringing down the last of the U.S. torpedo planes. . . . To the critical eye of the experts on the Striking Force, the American tactics were very primitive. Surprisingly, they had no fighter escorts. Nor did they split their attack, as the Japanese had learned to do. They foolishly kept in a single unit and simply hurled themselves at the *Soryu*. Bunched together, they were easy to shoot down. Their torpedoes (the men on the *Soryu* thought they saw four) passed harmlessly by the carrier."[36]

When U.S. Navy torpedo bomber pilots met Japanese pilots in battle, the Japanese simply pulverized them. This is the major theme in Kernan's 2005 book *The Unknown Battle of Midway: The Destruction of the American Torpedo Squadrons*. Kernan, a Midway veteran, makes it abundantly clear that for the badly trained, inexperienced, and poorly led U.S. Navy pilots, the Battle of Midway was an utter nightmare that should have ended in an unqualified and absolute defeat. The words "pusillanimous" and "unfit" frequently come to mind as one reads about the conduct and tactics of certain U.S. Navy officers in this "tell all" book. Despite knowing in advance that the Japanese were moving toward Midway for a showdown, and knowing the high stakes, the U.S. Navy pilots were incredibly unprepared. One torpedo plane pilot later noted that prior to this battle he had never even carried a torpedo, nor did he even know the *type* of torpedo that he would bring into battle against the Japanese! "So poor was their training," said Kernan, "that one pilot made a water landing with his wheels down and another landed downwind."[37] Most of the American torpedo planes were shot out of the sky by Japanese Zeroes, which was really easy to do, since most of the attacking U.S. torpedo squadrons did not have the fighter escort that they had been promised. This led some U.S. officers to accuse the fighter pilots of cowardice or incompetence, even though just months before the U.S. Navy had been bragging that its naval aviation units were "the best in the world."[38]

When U.S. torpedo planes did manage to survive long enough to launch their badly designed torpedoes, the results were inconsequential. "They went in separately, one squadron after another, on the morning of June 4, and all in all, fifty-one planes tried to hit the Japanese ships with torpedoes that day. Only seven landed back at base. This comes to an aircraft loss rate of over 86%. Out of 128 pilots and crew who were in torpedo planes that day, 29 survived, 99 died. And not one torpedo exploded against the hull of a Japanese ship."[39] Understandably, this made the Japanese "openly contemptuous of the Americans and their equipment."[40] Indeed, the American performance was so appallingly bad that day that some senior officers feared that even if they survived the battle, their careers would not; thus, says Kernan, a massive cover-up was orchestrated. Reports from the USS *Hornet* contained misleading remarks and omissions, leading Admiral Spruance to suggest that the *Hornet*'s report was less than completely candid about why so many of her airmen were killed that day while inflicting no damage on the enemy.[41] It truly was a miracle that the U.S. Navy even *survived* this battle, much less won it. As Tom Clancy tells it, the victory at Midway was a testament "to the Navy's skill and wisdom in deploying and fighting naval aviation."[42] This is absolutely preposterous and demonstrates that Clancy does not know the difference between "skill" and "plain old dumb luck."

By the way, if most American historians believe that Midway was the turning point in the war against Japan, not everyone else agrees. Admiral. S. G. Gorshkov, former commander in chief of the Soviet Navy, said that even after Midway,

> Japan's fleet retained superiority in forces, possessing (including those newly brought into commission) eight aircraft carriers as against four American. In battleships and cruisers the balance was also in favour of the Japanese. Even the character of the combat operations of both sides, following the Midway Island battle, indicates no turn in the course of the war. The Japanese continued to stage landings and conduct an offensive on New Guinea and the Solomon Islands and created a more serious position for the forces of the Allies by destroying two further American carriers *(Wasp* and *Hornet)*. Churchill wrote that in the 'autumn of 1942 the Americans . . . appealed . . . for one or more British carriers, . . . that an intense crisis had arisen in the Solomons.' In fact the Americans at that time were left with only two damaged carriers, the *Saratoga* and *Enterprise*. There was a real threat of invasion of Australia by the Japanese. What sort of turn in the course of the war was this?[43]

(And as we'll see shortly, the fact that the Americans had to ask the British for extra carriers tends to undermine the "superior U.S. industrial capacity won the war" argument put forth by many Americans.)

Some believe that the Japanese had a second chance to neutralize the U.S. Navy in the Pacific at Guadalcanal in September 1942. Orita conjectured that had the Japanese done what the Americans had expected at Guadalcanal—that is, confront the Americans with a vastly superior force, both morally and materially—the Americans would have lost the battle, the remaining U.S. carrier in the Pacific would have been destroyed, and Japan would have been unfettered and unrestrained in the Pacific for a very long time. Losing the carrier *Wasp* in September 1942 to the Japanese submarine *I-19*, the U.S. Navy had only a single operational carrier left in the Pacific, which had luckily avoided the torpedo that took her escort, the battleship *North Carolina*, out of action. During that month,

> One air strike and two submarine attacks had very nearly wrecked what part of the American fleet could be used against us. Now they had only one [carrier] left in the Pacific that could fight, USS *Hornet*. And only one battleship, USS *Washington*. . . . Against this single carrier in mid-September, our navy could range eight. While *Hornet* could put about 75 planes into the air against us, *Zuikaku, Shokaku, Zuiho, Taiyo, Hiyo, Unyo* and *Shoho* could launch more than 360, all told. Against *Washington* we could pit *Musashi* and *Yamato*, mightiest battleships ever built, plus eight other battleships far superior to the obsolescent ones America was keeping well to the rear. Mid-September of 1942 was the period of golden opportunity for the Combined Fleet.[44]

The Japanese were still much better trained than the Americans at that point, and it looked as though the failure to crush the U.S. fleet at Midway would be rectified in the Solomons.

Happily for the Americans, it was not to be. Instead, cautious Japanese officers did not use all the means at their disposal and consequently lost a great strategic victory. As Orita put it,

> We still had such superiority in forces that it seems almost unbelievable now that the chance to race down to Guadalcanal with overpowering strength was not seized. A swift and overwhelming blow could have been struck at Guadalcanal at any time between Sept. 15 and Oct. 1. There would have been absolutely no way for the Americans to counter it. . . . [I]n September, 1942, we had America nearly beaten in the Pacific. President Franklin Roosevelt at that time was actually considering whether or not to move his marines off Guadalcanal before they were

slaughtered. . . . Mr. Roosevelt was lucky. He put off making an immediate decision at all. Our high command solved his problem by *not* doing what Mr.Roosevelt feared most we would do—bringing down upon Guadalcanal all the force Japan could exert.[45]

Once again, a devastating strategic victory was denied to the Japanese, even though their forces were superior in most aspects, except that they did not have radar, but even this was not a fatal deficiency because as Overy said, "Against mass air attack even radar warning was of limited value."[46]

Lt. Col. Forrest R. Lindsey, USMC (Ret.), does not cover the possibilities of an American defeat at Guadalcanal, but he does agree that if the Americans had not been so lucky at Midway, the Japanese would have been "essentially unopposed from the Indian Ocean to the California coast."[47] The only thing standing in their way would have been the American submarine force, but in the early years of the war American submarines were severely handicapped by poor training, overly cautious skippers (many of whom were relieved of their commands—in fact Padfield said that proportionately more American submarine COs were fired than those of any other major navy),[48] and what Spector called the "worst torpedoes" in the world.[49] Harris uses the word "abysmal" to describe the performance of American submarines during the first two years of the war, and he backs up this assertion statistically: "The U.S. submarine score for 1942 was 180 ships, 725,000 tons (about equal to a monthly U-boat total). The Japanese replaced 635,000 tons in the same period."[50]

As history tells us, U.S. Navy torpedoes and submarine tactics improved markedly in the final years of the war, and the American submarine force played a decisive role in the Allied effort to beat Japan. Even so, Compton-Hall has argued that over the course of the war, British submarines were, boat for boat, generally more combat effective than American boats,[51] and German submarines seized at the end of the war were found to be technically superior to American boats in a number of ways. Orita also ventured that much of the success of American submarines in the waning years of the war was due to the facts that the U.S. Navy copied a torpedo developed by the Germans and that the American ASW forces had benefited enormously from a weapon they copied from the British, the Hedgehog.[52] It really is incredible that the American submarines did as well as they did in the final years of the war, because, in addition to the aforementioned shortcomings, and quite unlike the other major navies, U.S. Navy submarine skippers had to request permission over the radio before attacking. According to Granatstein, "Admiral Jeffry Brock recalled that the fleet signal book employed by the RCN. had one code for going into action: 'Enemy in Sight. Am Engaging.' The comparable USN code translated as 'Request Permission to Open Fire on the Enemy,' something that Brock was convinced was part and parcel 'of the determined resistance of American officers to make any move at all without the written and signed authorization of someone senior.'"[53]

Had the Americans lost at Midway, the possible consequences for the U.S. Navy could have been rather substantial. Lindsey projects that the Japanese could have moved on to capture Hawaii and then proceeded against the American mainland: "Japan's enormous striking power could reach and severely damage the cities, factories, transportation, and fuel reserves on America's west coast. Strong enough attacks would also convince America's leaders that continued war against Japan was impossible. . . . The major American aircraft companies were well within carrier-based aircraft range and some were even within range of [Japan's] battleship's guns from fire support areas along the Pacific coast."[54] If this had happened, and it certainly was a strong possibility, then the modern-day "big carrier" U.S. Navy might have evolved quite differently, to say the least. Orita suggested that the Japanese submarine force was actually quite successful in the Indian Ocean, with one of its submarines destroying thirteen enemy ships, totaling 78,000 tons: "[Commander] Fukumura got 9 of these—an excellent example of what the Japanese 6th Fleet might have accomplished had the Battle of Midway been won by us and all our other submarines loosed for attack operations in the west. Australia and India would have been cut off by sea. Years might have passed before any kind of major offensive could have been mounted against Japan, if at all!"[55] Imagine what it would have been like if only the Japanese had:

- Coordinated with the Germans (who easily could and should have continued their U-boat campaign in American waters for at least a full year), and
- Won at Midway, as they easily should have.

As suggested earlier, both coasts of the United States would have been subject to intensive and relentless attacks simultaneously during the second half of 1942, with German U-boats destroying oil tankers in the Atlantic and Japanese battleships and carriers off the coast of California, blasting San Diego, Los Angeles, and San Francisco, all the while facing little resistance from the decimated and emasculated U.S. Navy. Imagine the carnage.

Some opine that America would have won the war in the Pacific easily anyway, simply because it was able to "outproduce" Japan, or, as one apologist said recently, "From January 1942 to August 1945, the United States launched 37 fleet carriers, 83 escort carriers and 349 destroyers. The Japanese built three fleet carriers, six small-carrier conversions, and 63 destroyers. Even if those sneaky, treacherous [Japanese] could have destroyed 50 percent of the West Coast production facilities, the war effort would not have been slowed, much less crippled."[56]

Case closed? Well, not quite. Lt. Burdick Brittin, USN, a Midway veteran, confided to his diary in the final days before the battle, "We have history in the palm of our hands during the next week or so. If we are able to keep our presence unknown to the enemy and surprise them with a vicious attack on their carriers, the U.S. Navy should once again be supreme in the Pacific. But if the [Japanese] see us first and attack us with their overwhelming number of planes, knock us out of the picture, and then walk in to take Midway, Pearl Harbor will be almost neutralized and in dire danger—I can say no more—there is too much tension within me—the fate of our nation is in our hands."[57] Apparently, Brittin was not fully convinced that the superior industrial capacity of the United States would make any difference, since it would clearly take much time to recover, rebuild, train, and deploy a new fleet to replace one that had been obliterated. This young officer was afraid, and for very good reason.

As Shuger said in 1988, "Even the briefest review of military history also reveals that for every battle decided by superiority of weapons, there are ten in which the outcome depended on differences in intelligence, planning, tactics, communications, logistics, or resolve."[58] Indeed, if simply building more (and technologically more advanced) ships, tanks, and airplanes than your enemy, in and of itself, were a guarantee of an easy or inevitable victory, then how could Afghanistan, one of the poorest countries in the world, repel the gargantuan Soviet armed forces in the 1980s? Why is the United States still there fighting the remnants of the Taliban? How could North Vietnam have endured the most ferocious air assault in history long enough to force the world's richest country to withdraw from South Vietnam? (During the air war over Southeast Asia, the United States dropped the conventional equivalent of 640 "Hiroshima-type" atomic bombs, yet it did not win.[59] Vietnam, Laos, and Cambodia fell to the communists.) Sadly for the Americans, we must discredit the commonly held suppositions of people like the late Adm. Arleigh Burke, USN, who in 1964 asked, "Do we really believe that a nation that's starving can field a more powerful force in South Vietnam than we—the most powerful nation in the world?"[60] Physically, definitely not, but morally, yes, absolutely. As Hammes tells us, "In Vietnam, the side with overwhelming wealth, power, and technology [was] decisively defeated."[61]

Moreover, how does Israel, badly outnumbered by its neighbors, manage to survive, let alone be the dominant military power in the Middle East? As Dixon put it, "There is the Israeli Army, the David of two and a half million Jews who in six days defeated the Goliath of 100 million Arabs. By its competence and vastly superior direction this miniscule army, drawn from a country poor in resources and gravely disadvantaged by its geographical position, managed to defeat an enemy from countries possessing inexhaustible reserves of natural wealth (including one half of the world's hydrocarbon reserves)."[62]

How, too, were the Finns able to vanquish the Russians in 1939–1940?[63] How on Earth could resource-poor Japan squarely defeat Russia in 1905 (as Prange so eloquently pointed out, "More than once in sea warfare a great sailor has snatched victory from a superior foe")[64] and then invade and occupy China in the 1930s? Going farther back into history, how was it possible for Hannibal and the Carthaginians to rout the much superior Roman army at the Battle of Cannae,[65] for Archimedes to single-handedly keep a Roman fleet from taking Syracuse for over a year,[66] or for Napoleon to clobber much larger Austrian and Russian forces at Austerlitz in 1805?[67] Or, for the Korean navy under Admiral Yi Sun-Sin to defeat much larger Japanese invasion fleets in 1592 and 1598, in the latter case killing some forty thousand Japanese warriors?[68] And how did Rear Admiral Pavel Dzhones, fighting for Russia, defeat larger Turkish naval forces on the Black Sea in 1788?[69] War production is only one factor among many in the combat equation, and it is frequently a rather misleading one at that. Biddle declares in 2004 that contrary to popular opinion, "predominance," as measured by military expenditures, war materiel, and the number of personnel, is, as an independent variable, a very poor predictor of victory: "Real battle outcomes cannot be explained by materiel alone; in fact, materiel factors are only weakly related to historical patterns of victory and defeat."[70] Using sophisticated mathematical models, Biddle has demonstrated that the outcomes of combat in the twentieth century clearly undercut the outdated notion that "bigger and more expensive are better" in battle. As he said, "All told, the data show no support for a simple assumption that preponderance predetermines capability."[71]

Of course, there are many who persist in believing that superior industrial production produces victory, and I suggest they keep the following in mind. Some have actually cited automobile production statistics, for example, to suggest that Japan could not possibly have beaten the United States in World War II, which they see as a war fought by factory workers as much as warriors (as Lord said, "in 1940 the United States turned out 4,500,000 automobiles, while Japan only made 48,000").[72] With all due respect to Admiral Yamamoto and many others, I do not see how prewar automobile production is relevant at all: the Imperial Japanese Navy was technically more powerful than the U.S. Pacific Fleet at the beginning, and certainly better trained, and even though its leaders made plenty of mistakes, it proved to be a treacherous adversary for four bloody years. While the Americans were building all these cars, the Japanese were concentrating on building the best aircraft carriers, fighter planes, torpedoes, and battleships in the world. I would also point out that in 2004 the United States produced just 9.6 percent of the world's automobiles, while Japan came up with 19.7 percent and Germany 11.7 percent.[73] Not that I think it is terribly relevant, but for those who do, the industrial output ratio has indeed changed, and not in America's favor.

War production is a salient factor but not the sole and direct reason for victory, any more than cows and pigs are the causes of obesity or automobile manufacturers are the sole causes of car accidents. Think Mogadishu or *Black Hawk Down* for a recent example. And in 1998, Greider pointed out that "a still-classified study by the Defense Science Board concludes that a regional adversary, by spending $10 billion a year on defense and such things as missiles, commercial space satellites, and hardened underground facilities, could insulate itself against a U.S. invasion. 'They could really screw up our current forces,' Vickers concedes."[74] (Vickers was the director of strategic studies at the Center for Strategic and Budgetary Assessments in Washington.) *This is a key point, one that needs to be kept in mind—that just because the U.S. Navy has no direct challenger right now does not mean that a smaller enemy who knows how to exploit that service's weaknesses cannot defeat it.* As retired Philippine army brigadier general Victor N. Corpus said recently, "Asymmetric warfare may be compared to a fierce lion invading the territory of a school of piranhas; or a king cobra encroaching into a colony of fire ants. The lion may be the king of beasts, mighty and strong, but it is no match against the tiny piranhas in their own territory. The sharp fangs and claws of the lion are rendered useless. The same is true with the cobra's venom. The analogy applies to the French in Dien Bien Phu, the Soviets in Afghanistan and the Americans in Vietnam and now in Iraq."[75]

When in doubt, always remember the immortal words of Mark Twain: "It's not the size of the dog in the fight, it's the size of the fight in the dog."[76] It could be argued that even if Midway had ended in its favor Japan would have been ultimately defeated, but not for the reasons commonly supposed. Overy argues convincingly that Japan's defeat had more to do with the loss of its warrior spirit in its long and wearing war with China, which had begun in 1931, and with the home front rather than any other single factor.[77] In other words, the Japanese public, not so much the ordinary soldiers, sailors, marines, and airmen themselves, many of whom were quite willing to sacrifice themselves to protect the emperor, was simply tired of making war and was not truly stalwart when things began to sour after the loss at Midway. A few of the most senior military leaders too, such as Admiral Yamamoto, had, shall we say, "defeatist" tendencies from the very beginning, which probably did not help. Yamamoto did not realize that, as demonstrated here, history is full of examples of the weak beating the strong.

Reluctance, diffidence, aboulomania, lack of coordination with the Germans, and insecurity also robbed the Japanese of many additional opportunities for victory, and that right from beginning, with the abbreviated attack on Pearl Harbor. The prevailing wisdom in Western circles is that the attack on Pearl Harbor was a mistake, but this theory has its critics. Russett argues that the attack on Pearl Harbor was rational, well planned, and, from a Japanese standpoint, quite necessary and unassailable as a counter to attempts by Washington and its allies to cut off Japan from its overseas natural resources.[78] The attack itself was well executed, but it did not go far enough, and the Japanese did not press on when they clearly had all the means to do so. Hoyt criticizes the Japanese for putting too much emphasis on hitting the American battleships at the expense of easier and, possibly, more strategically valuable targets. "So eager were the Japanese fliers to sink battleships," he assesses, that "they ignored the tanker *Neosho*, which was loaded with high-octane aviation fuel, If they set her afire she might have burned down the whole harbor,"[79] thus denying the Americans the use of one of their most important bases.

"The more important error," Hoyt maintains, "was the failure of the Japanese to cripple the Pearl Harbor submarine base, which they could have easily done with another attack.... Also, four-and-a-half million barrels of oil had been stockpiled at Pearl Harbor, located in dumps above ground, made an easy target. The Japanese ignored them."[80] All of these opportunities were extinguished simply because Admiral Nagumo lost his nerve and halted the thoroughly one-sided Battle of Pearl Harbor much too soon, when his enemy was very much at a disadvantage. (According to van der Vat, Nagumo "was prone to bouts of anxiety which prevented him from sleeping; even the smallest decision caused him stress.")[81] An extra day of attacks would have been all that was necessary to put the whole base and its ships out of action, or even out of existence, which in turn would have made life much easier for the Japanese in the years ahead. Such an extension might have allowed the Japanese to destroy the returning carrier USS *Enterprise*, which would have been back at Pearl Harbor by December 6 if not for delays caused by bad weather. If this had happened, or if the *Enterprise* had not been delayed, her likely demise might have had a dramatic effect on the U.S. Navy during the Battle of Midway, where she was one of the major contributors. As military historian Elihu Rose has written, "In a tantalizing what if, one might speculate upon the outcome of the battle [of Midway] had not the *Enterprise* been out of harm's way on December 7."[82]

And finally, while there is no doubt that, as it happened, America's carrier task forces and submarines played decisive and integral roles in the eventual defeat of Japan, many Americans overlook the significant contribution of the Soviet Union to that same end. In actuality, the Red Army was responsible for neutralizing approximately 32 percent of Japan's army, but this fact seldom appears in the typical American discourse on the war.[83] As Admiral Gorshkov suggested, "Victory over [Japan] demanded a sustained struggle with the enlistment of large land forces, which the allies did not have.... That is why the entry of the U.S.S.R. into the war against Japan was so necessary for the allies. Without it, it was not possible to break the determination of the Japanese militarists to continue the war.... It is known that in fulfilling their obligations as allies the Soviet Army and Navy with a powerful blow smashed the Kwantung army and the Japanese support points on Sakhalin and the Kurile islands, after which Japan was forced to surrender unconditionally."[84]

6

The Russians Mug the *Kitty Hawk,* the *Saratoga,* the *Constellation,* the *Carl Vinson,* and Others

> If there was any doubt about Soviet intentions . . . one had only to read the speeches of the Soviet naval commander, Adm. Sergei Gorshkov, who had boasted that the United States had made a strategic miscalculation in relying on large and increasingly vulnerable aircraft carriers to project power in the world. The U.S. strategy would fail in wartime, Gorshkov alleged, because "the combat potential . . . of nuclear-powered aircraft carriers is inferior to the strike potentials of submarine and air forces."
>
> —PATRICK TYLER[1]

THE EXAMPLES ABOVE FROM unscripted naval exercise evolutions provide ample evidence of the vulnerability of U.S. Navy carrier battle groups to attacks from diesel submarines, but of course there are other ways to sink a carrier, as the Russian air force knows well. In October 2000, the smart-looking aircraft carrier USS *Kitty Hawk* was "mugged" by Russian Su-24 and Su-27 aircraft that were not detected until they were virtually on top of the carrier. The Russian aircraft buzzed the carrier's flight deck and caught the ship completely unprepared. To add insult to injury, the Russians took very detailed photos of the *Kitty Hawk*'s flight deck and, very courteously, provided the pictures to the American skipper via e-mail. In a story in the December 7, 2000, edition of *WorldNetDaily,* one U.S. sailor

The Russians mugged USS *Kitty Hawk* (CV 63) in 2000. —*S. Rowe, U.S. Navy photo*

exclaimed, "The entire crew watched overhead as the Russians made a mockery of our feeble attempt of intercepting them."[2] Russia's air force is now only a faint shadow of what it once was, but even now it can demonstrate that it can, if necessary, do significant damage to the U.S. Navy. It is little wonder then that a Russian newspaper gloated, "If these had been planes on a war mission, the aircraft carrier would definitely have been sunk."[3]

Perhaps they are right. As Howard Bloom and Dianne Star Petryk-Bloom advised in 2003, both the Russians and Chinese now have the deadly SS-N-22 Sunburn missile at their disposal. This massive long-range missile, equipped with nuclear or conventional warheads, is extremely difficult to detect or destroy. According to Jane's Information Group, it is more than capable of destroying any U.S. aircraft carrier.[4] More to the point, Timperlake (a former U.S. Marine Corps fighter pilot and Naval Academy graduate) and Triplett warned that the Sunburn missile is "designed to do one thing: kill American aircraft carriers and *Aegis*-class cruisers. The SS-N-22 missile skims the surface of the water at two-and-a-half times the speed of sound, until just before impact, when it lifts up and then heads straight down into the target's deck. Its two-hundred-kiloton nuclear warhead has almost twenty times the explosive power of the atomic bombs dropped on Hiroshima. . . . The U.S. Navy has no defense against this missile system. . . . As retired Rear Admiral Eric McVadon put it, 'It's enough to make the U.S. 7th [Pacific] Fleet think twice.'"[5] This new weapon, if it is ever used against the U.S. Navy, might destroy not only a large number of American ships but also Tom Clancy's rather interesting claim that "today's [carrier battle group] tactics revolve around the reality that in the post–Cold War world very little threatens U.S. naval forces."[6]

Some would say that this example does not validate the anticarrier argument, because in a real war the carrier and her escorts would have been more careful and at a higher level of readiness. Indeed yes, but what if this mock attack had been the opening shot in an unexpected war? In that case, the arrogant and myopic thinking of the U.S. Navy probably would have cost it one multibillion-dollar carrier and probably some of its escorts on the very first day. Multiple coordinated surprise attacks by aircraft, cruise missiles, and diesel submarines could quickly emasculate many of the American carrier battle groups. And I feel obligated to point out that even if the carrier's Aegis-equipped escorts had been on high alert, and indeed been running with their radars at high power, this could have made the group vulnerable to Russian antiradar missiles. Holland noted in 1997, "Ironically, the highly sophisticated computer and strong radar systems that compose Aegis also make an Aegis carrying ship an easily recognizable target."[7] And as one naval aviator told Wilson during his visit to the USS *John F. Kennedy*, the Russians "can make it rain longer than we can swim."[8]

That is a politically incorrect statement for a naval officer, to be sure, but others have gone farther. Capt. T. S. Teague, USN, broke one of the cardinal rules of the U.S. Navy when he, the skipper of the *Kitty Hawk* in the early 1980s, told Stevenson that yes, the Russians could "take out" his ship if they made an effort, and this was long before the Russians developed the SS-N-22.[9] For some reason, possibly convenience or wishful thinking, many U.S. analysts tend to overlook or downplay the fact that the Soviets had deployed submarines with nuclear-tipped torpedoes more than forty years ago, and if even a few dozen of these weapons could be used effectively, the surface forces of the U.S. Navy could be incinerated in short order. This is not new technology at all, and it wise to predict that eventually these weapons will fall into the hands of many nations, some of which might wish to oppose the United States. It would be apt to say that not only *can* U.S. Navy carriers be destroyed, as evidenced by combat actions involving various battleships and big carriers in World War II and the frank admissions of U.S. Navy officers, they can *definitely be destroyed* by a determined enemy, with good diesel submarines, good crews, and good torpedoes or cruise missiles.

Also, supercarriers can even be rendered harmless, at least temporarily, by things far less impressive than cruise missile or torpedo attacks, nuclear or otherwise. In 1975, for example, the *Kennedy* was rendered dead in the water for four hours, and therefore almost useless and extremely vulnerable to a potential adversary, and all as the result of a collision with one of her escorts. According to Vistica, the CNO, Adm. James Holloway, USN, was very concerned about this, as it contradicted his own statements that nothing short of a nuclear weapon could stop a supercarrier.[10]

Getting back to the *Kitty Hawk* incident, a Navy spokesman said that the *Kitty Hawk* had not been surprised, that they had known the Russian planes were not going to attack and had tracked them almost from the moment they took off. In other words, "We were on top of things, no need to intercept, and certainly no reason for alarm." When the Russians overflew the *Kitty Hawk*, the carrier was "taking on fuel, it was not sailing fast enough to launch its aircraft."[11] It took forty minutes for the first American aircraft to be launched, and the Russian air force was delighted with the results: "'For the Americans, our planes were a complete surprise,' said General Anatoly M. Kornukov, the Russian air force's commander in chief. 'In the pictures, you can clearly see the panic on deck.'"[12] This episode sounds somewhat like what happened to the Imperial Japanese Navy at the Battle of Midway, where its aircraft carriers were caught off guard and attacked while their planes were being rearmed. Clever enemies often prefer to attack during periods of low readiness, or during poor weather.

Just for the sake of argument, let us assume that the U.S. Navy had indeed tracked the Russian planes and fired at them (and/or their attacking cruise missiles). Even if this had happened, it still does not mean that the crafty Russian attack would have failed. The reasons for my pessimism were contained in a foreboding 2000 report by the U.S. General Accounting Office. The report cast great doubt on the survivability of American surface ships, because the Navy had continuously exaggerated "the actual and projected capabilities of surface ships to protect themselves from cruise missiles because the models used in the assessment . . . include a number of optimistic assumptions that may not reflect the reality of normal fleet operations."[13] For example, in its highly questionable testing of shipborne defensive systems, the U.S. Navy assumed any such attack would occur in perfect weather, with a perfect American crew, and flawless equipment, which is a highly unlikely scenario.¹⁴ This, most certainly, is a theoretical argument, but if one needs a real-life example, recall the nasty business of 1987 in which an Iraqi pilot, flying a French-designed Mirage F-1 fighter, fired two Exocet missiles at the USS *Stark* and nearly sunk it. Although the Iraqi aircraft was detected, *none of the U.S. missile warning systems on the* Stark, *the nearby USS* La Salle, *the USS* Coontz, *or an Air Force AWACS plane in the area detected the incoming missiles, which struck the ship and killed thirty-seven men. Not one of these systems did what they were designed to do, and that was to protect the ship from attack.* As one senior officer on the *Stark* put it: "The sensors, in fact, did not work as advertised. The launch was not detected on any of the four search radars, the SLQ-32 did not correctly identify the threat [and] the AWACS did not detect the launch."¹⁵

Even if there had been F-14s on combat air patrol (CAP) above the carrier during the Russian penetration, their outrageously expensive Phoenix or Sparrow missiles might not have made any difference, either. According to a 2001 paper by Col. Everest Riccioni, USAF (Ret.), "The long range U.S. Navy Phoenix missile was fired twice in combat in 30 years and missed both times—a zero return on a large investment."¹⁶ (Note that Riccioni was referring only to the U.S. Navy; some claim the Iranian F-14/Phoenix units were more successful.) Riccioni's research indicated a clear inverse correlation between expense and the probability of getting a kill, or to put it another way, the more expensive the missile the less reliable it is. The Phoenix missile (now retired) was easily the most expensive air-to-air missile in the world, and it was much less reliable than cheaper missiles or guns.¹⁷ As Chester Richards argued, "In fact it would not be facetious to suggest a new law of combat effectiveness: The side with the most expensive weapons loses."¹⁸

The U.S. Navy never properly or realistically tested the Phoenix, either. Said Fallows,

> Because of prohibitive costs, we have never conducted realistic operational training with the F-14 firing Phoenix missiles in the presence of jamming and tactical countermeasures. [Secretary of Defense Harold] Brown was saying that the biggest question about sophisticated, precision-guided weapons—whether they can overcome the efforts any competent adversary would make to thwart them, by jamming or deception or anything else—has never been answered in realistic practice). Nor have we demonstrated that we can load and launch the large number of

Phoenix missiles against multiple targets that would be required to defense against a determined ... attack."[19]

That was written in 1981, but those nagging questions remain even today. Thank goodness the United States has chosen only incompetent enemies since Vietnam. Some say the Iranians, however, may have done more testing on the Phoenix; if so, and if their claims of success against the Iraqi air force are true, the U.S. Navy should be deeply embarrassed that another country bought a U.S. weapon system, tested it more thoroughly, and may have used it more successfully than it did. Still, as even Cooper and Bishop admit, "It remains unclear exactly how many air-to-air kills were scored by IRIAF [Islamic Republic of Iran Air Force] F-14s between 7 September 1980 and 7 July 1988, as Air Force records were repeatedly tampered with during and after the war, mainly for political, religious or personal reasons. This has led to considerable confusion."[20]

David Isenberg was not enthusiastic about the Phoenix, either. In 1990, he wrote, "The Phoenix has long been plagued by design and mechanical defects.... Even if Phoenix missiles were consistently free of defects, they would still have several serious disadvantages. Radar-guided missiles such as the Phoenix emit powerful electronic waves that enemy weapons can home in on, a feature that makes targets out of the ships, planes and artillery units that fire those missiles."[21] As we will see later on, the enemy can exploit these powerful radar emissions. Additionally, said Shuger, "The 1981 shootdown of two Libyan Su-22's by two F-14s was hardly a significant test. It was a mismatch if there ever was one: Leading edge fighters with fully exercised crews against clumsier ground attack aircraft flown by pilots so nervous they fired their missiles well beyond range."[22]

Note that although this encounter between the U.S. Navy and the Libyan air force was a success, there have been other encounters in which the Libyans almost turned the tables. Robert Wilcox reports that one day in 1981, two F-14s from the USS *Nimitz* were returning from their CAP station and "had almost been sneak-shot by two fast-moving MiG-23s."[23] Since things had been quiet that morning over the Gulf of Sidra, the Navy pilots "decided nothing was going to happen.... To conserve fuel, they'd exited slow. That had been a mistake. With their backs turned, the Libyans, seeing a chance for a hit, had come after them—and at a much higher speed. It was a smart tactic. The distinctly swing-winged 'Floggers,' as NATO designates the MiG-23s, had heat-seekers, and the Tomcats were showing red-hot exhaust. Luckily, the relief CAP, two F-14s from VF-84, the *Nimitz*'s other fighter squadron, spotted the trailing MiGs on radar before they (the MiGs) were in shooting range, and gave warning."[24] This saved the day for the Americans, but one has to wonder what might have happened had the relief CAP aircraft been delayed for even a few seconds or if their radar had malfunctioned, as the MiGs had apparently closed to within three miles and had been undetected by the returning F-14s. "'Not a great spot to see your first MiG,' later wrote Snodgrass, '3 miles in trail at 6 o'clock.' But he and Kleeman had relearned a lesson: never relax until you're safely back on ship."[25]

If there had indeed been a fight in the air that day in 2000, we should keep in mind two things. Firstly, there was always a small chance that a U.S. missile might have homed in on the *Kitty Hawk* herself (these things can happen). Secondly, in Vietnam, Soviet AA-2 air-to-air missiles actually had a higher "probability of kill" per launch (about 22 percent) than the three most commonly used U.S. missiles of the time (averaging about 11.2 percent).[26] And despite the lopsided air-combat maneuvering (ACM) success of the U.S. Air Force (not the Navy) in Operation Desert Storm (during which it was obvious that Saddam Hussein and many of the fleeing Iraqi pilots had neither the courage, training, nor tenacity of a thoroughly committed opponent, like the Japanese kamikazes or today's Islamic extremists, for example), it might not be a good idea to take for granted that today's U.S. air-to-air missiles are more reliable than their Russian counterparts. Indeed, many say a recent Russian missile, the AA-12 Adder, is comparable or even superior to the U.S. AIM-120 AMRAAM.[27] Many fear AMRAAM (Advanced Medium-Range Air-to-Air Missile), but there are certainly ways to deal with it. The British, for example, have learned how to shoot down AMRAAM-equipped aircraft without suffering any losses themselves, at Exercise Maple Flag in the late 1990s.[28]

Another question now comes to mind. If the crew of the *Kitty Hawk* really knew of the impending Russian visit, why did the U.S. Navy decline to release the Russian photos? If the crew had truly not been surprised, the photos of the flight deck should surely reveal this and clear the U.S. Navy. If there had been some classified equipment or activity depicted in the Russian photos, surely the Pentagon could have censored the photos as required, then released them to show the world a crew at sea going about routine business.

Why also did the *Kitty Hawk*, forty minutes later, finally launch aircraft to intercept the Russian planes that had already flown over but done no physical harm to the ship? Why was it necessary belatedly to intercept the Russians if the U.S. Navy was so confident that the Russians were no threat? And why did the *Washington Times* impart that the "*Kitty Hawk* commanders were so unnerved by the aerial penetration they rotated squadrons on 24-hour alert and had planes routinely meet or intercept various aircraft"?[29] Because in asymmetrical warfare, the very concept is to strike when the larger, more powerful enemy is least prepared. This is what the Japanese did when they attacked Pearl Harbor in the early morning hours on a Sunday. This is why the 1968 Tet holiday offensive was launched when the Army of the Republic of Vietnam was in a low state of readiness. But then, perhaps it would have been more sporting of the Russians to have called in first before launching their mock attack.

As an aside, although the foregoing concerns fast jet aircraft, the U.S. Navy has had troubles dealing with slower planes as well. Shuger told an interesting story in his 1988 unpublished book manuscript *Navy Yes, Navy No*:

> The Navy has established a classified radius X nautical miles around a carrier within which all aircraft will be escorted by airwing planes. As you might expect, not *every* intercept is perfect, so there are occasions when the bogey isn't intercepted until less than X. (I know of at least one occasion in 1981 during the hostage crisis, where an Iranian P-3 patrol craft—which conceivably could have been carrying U.S.-made Harpoon antiship missiles—flew more or less undetected and unintercepted within visual range of a carrier.) But fleet message writers feel that their job security requires that no reports to that effect ever leave the ship. So there is a definite party line about intercepts: They *always* take place at X nautical miles.[30]

A colleague recently reported a very similar incident between a Royal Australian Air Force (RAAF) P-3C Orion and an American aircraft carrier, also in the early 1980s. In the words of (Ret.) Squadron Leader J. R. Sampson, RAAF:

> When I was an RAAF liaison/briefing officer en route from Diego to Perth for R&R sometime in 1981/82, I dined in the [American] admiral's suite and the admiral gave me a copy of a message that censured an air wing commander for allowing an RAAF P-3C to get in undetected amongst the CVBG [carrier battle group] screen a few days earlier.

According to the message the commander himself was in an F-14 cockpit checking out the TCS [television camera set] that had just been installed as a new piece of F-14 kit. TCS enables long-range visual identification of targets. He was adjusting the FOV [field of view] when he saw a P-3 swim across his screen, right on the carrier's bow at about 300 feet above sea level. He'd just come from CIC [combat information center] and knew that no cooperating P-3's were due so he queried the FLYCO who queried the CIC who asked the on station E-2C. They didn't even have the capability to launch an F-14 intercept. Very embarrassing but the admiral gave me a copy of the message to take back to headquarters[31]

Embarrassing yes, and it proves that an enemy doesn't even need speedy jet fighters to get through a U.S. Navy battle group's defenses. A large and relatively slow turboprop aircraft like the P-3 can do it just as well.

It almost goes without saying that even the older and relatively noisy Soviet/Russian submarines have a long tradition of tracking and stalking American carriers, but the American public occasionally needs reminding. The Soviets maintained a huge force of both nuclear and diesel submarines, and their boats were able to locate, pursue, and close with U.S. Navy carrier battle groups on many occasions. In 1966, the noisy Victor-class nuclear submarine *K-181* trailed the carrier *Saratoga* and her escorts in the Atlantic for several days and made "nine simulated conventional torpedo attacks on the aircraft carrier, from different directions and distances, and sent twenty radiograms on the task group actions to fleet headquarters. The *K-181*'s expert radiomen recorded the sound of the aircraft carrier's turbines at different depths, invaluable information for another cruise."[32] Although the Soviet submarine was eventually detected, it was not by the carrier or by her escorts but by the SOSUS warning net. Regardless, "it was a considerable triumph to put *K-181* within killing distance of the aircraft carrier."[33] Although the Americans did detect the *K-181,* it was not until well after she conducted her simulated torpedo attacks. In a real war, the carrier probably would have been destroyed before the Soviet sub could have been localized and attacked.

Of course, the U.S. Navy did the same thing to the Soviets as well, but most of us in the West either do not know or do not want to know the other side of the story. Something very similar happened in late 1967, said Tyler. A U.S. carrier task force in the Atlantic "had been shocked by the sudden appearance of the conning tower of a Victor-class submarine. The

Russian had popped up to thumb his nose at the Americans and to demonstrate a Soviet capability to penetrate the carrier battle group. It was a secret and unreported victory for the Soviet Union and an embarrassing and ominous moment for the U.S. Navy."[34]

Moreover, according to Kenneth Sewell, in 1968 a possibly rogue Soviet ballistic-missile submarine, the *K-129,* may have made an attempt to approach Hawaii. Dr. John Craven, former chief scientist at the U.S. Navy's Special Projects Office, has indicated through a "Bayes subjective probability assessment . . . that there was a small probability that the Russian submarine, whatever its name and class, could have reached Hawaii with a nuclear warhead."[35] SOSUS, as one retired senior submariner with direct experience stated, was far from wonderful in the 1960s, especially on the West Coast:

> In my 3 years on the West Coast net, I am not aware of a single verified detection of a single Soviet submarine from a West Coast SOSUS station, much less a two-station fix or a station-to-station tracking. We did succeed nicely against the noisy U.S. nukes of the era. I personally participated in the tracking of *Nautilus* from leaving port to the Aleutians on her first try to circumnavigate the Pole. Also tracked *Skate* from Pearl to the Aleutians. If a diesel boat lit off near the array, we got it, but when it shut down or when the range got more than a couple hundred miles, we had nothing. SOSUS was a great idea and did provide some good detections as time went on, but when I was in it, we groped for intel and found very little.

Things were not great in the late 1980s, either, said Hayes et al. "While U.S. intelligence capabilities are almost omnipresent in the Pacific, they are also unreliable. Indeed, U.S. intelligence systems in the Pacific may be so vast as to be unmanageable, and the intelligence itself so voluminous as to be incomprehensible. 'Intelligence' may become assured ignorance. Automated and bureaucratic systems are notoriously fallible, and Pacific Command intelligence is both highly automated and highly bureaucratic."[36] This might allow enemy submarines to slip through undetected, unchallenged, and therefore, undeterred.

Shuger also wrote in the late 1980s that the U.S. Navy sometimes had great difficulty locating even the oldest Soviet submarines: "More than a decade ago, when there were dozens of U.S. ships and planes in the South China Sea looking for the one Soviet submarine then on patrol there—and it was obsolescent one at that; this was before the big Soviet naval build-up of the Cam Ranh Bay naval base in Vietnam—it would still go unlocated for weeks on end. Imagine the difficulties presented nowadays by the increased numbers of quieter Soviet subs."[37]

Even the ill-fated Russian submarine *Kursk* gave the Americans a good run for their money. Truscott noted that in 1999,

> According to the Russians, the U.S. spent tens of millions of dollars trying to track the *Kursk* in the Med, with mixed success. The U.S. Sixth Fleet, based at Naples, became extremely active in the search for the *Kursk*. By Russian accounts, the U.S. Sixth Fleet restricted operation of its large ships and aircraft carriers to stay out of the Russian sub's possible area of activity. Captain First Rank Lyachin was certainly proud of the *Kursk*'s performance, later saying that the boat received a lot of attention from NATO's subs, ships and planes, but 'we almost always spotted them first.' NATO found it difficult to establish prolonged surveillance and contact with the *Kursk*.[38]

Given the U.S. Navy's tradition of substandard ASW, Captain Lyachin's claims are not difficult to believe.

Specific encounters between Soviet/Russian subs and American ships are rarely publicized or described in so much detail, but Sontag et al. have related that during the Cold War "Soviet subs seemed to be waiting to monitor U.S. naval exercises even before U.S. ships and subs arrived on site. A few times, Soviet subs had shown up in waters where U.S. exercises had been scheduled, then cancelled. Other times, Soviet subs barreled right into the middle of exercises almost as if they were trying to see how the U.S. forces would react."[39] "In 1985," said Weir and Boyne, "the Soviet submarine *K-324*, taking advantage of temperature variations in the Gulf Stream, detected American SSBNs (nuclear-powered ballistic missile submarines) on three different occasions, maintaining a combined contact time of twenty-eight hours," while another Soviet nuclear boat surreptitiously tailed another American SSBN for five days.[40] In these cases, we can see that the U.S. Navy's traditional belief in the inherent superiority of American nuclear submarines and tactics over their Soviet/Russian adversaries is not always justifiable or realistic.

Along those same lines, Kaylor reported that in 1986 "a U.S. attack-submarine skipper received a shock while tracking a Soviet sub in its home waters. The U.S. commander expected to hear his quarry long before it heard him. But suddenly the American sonar men heard a single loud metallic 'ping' in their earphones. Listening with passive sensors that make no sound of their own the Soviets had picked up the American sub, then transmitted a single 'active' sound wave to fix its exact location. It was the Soviet captain's way of saying, 'Gotcha!' But in wartime, it would have been followed swiftly by a torpedo."[41] This incident sounds very much like the one reported by Capt. Peter Huchthausen, USN, and his coauthor, Captain First Rank Igor Kurdin, Soviet Navy, in which a very noisy Yankee-class ballistic-missile submarine, the *K-219*, was detected by SOSUS as it approached the U.S. east coast that same year and was intercepted by attack submarine USS *Augusta*. The technically outmatched Soviet boat had a good skipper, though, and he found a way to get the jump on the American boat by hiding in a thermal layer, and then used its active sonar to ping the *Augusta* at very close range. In so doing, the noisy Soviet sub could easily have attained a firing solution before the American boat could have, and therefore probably would have gotten in the first shot if actual combat had ensued.[42] Some have declared that between 1965 and 1975, "there had been more than 110 possible detections of U.S. surveillance subs actively operating against the Soviet Union."[43]

A Canadian warship also surprised and pinged a very modern *Ohio*-class SSBN, the USS *Michigan*, during filming of the informative 1992 NOVA documentary "Submarine." This, of course, flies right in the face of the U.S. Navy's standard ballyhoo that non-U.S. forces cannot detect its SSBNs. A U.S. nuclear submariner, who wished not to be identified, also told me recently about an incident in which his elderly and relatively noisy nuclear attack submarine (launched in the early 1960s) had stalked and launched a simulated attack on a supposedly undetectable *Ohio*-class SSBN using passive sonar, "because our sonar chief steered us towards a part of ocean with lower than expected background noise, from the boomer screening the ambient noise. We were able to creep into their baffles to fire a water slug."

In addition, the American media learned in September 1997 that a Russian nuclear submarine had gotten uncomfortably close to the carrier USS *Constellation* and other ships during a Pacific cruise—so close, in fact, an anonymous U.S. Navy source "concluded later that the submarine would have sunk the *Constellation* near Seattle if there had been a conflict."[44] And Gertz recorded that "The submarine . . . loitered off the Washington coastline and practiced attack operations against the [USS] *Carl Vinson* during the carrier's training mission."[45] He concurred that the U.S.

Navy had great difficulty tracking the elusive submarine, and as he put it, "the Russian Oscar II–class guided missile submarine spent nearly two weeks in September about 100 miles off the Washington coastline and sailed undetected for days, eluding U.S. surveillance vessels and aircraft."[46] And even though their nuclear boats were very noisy until sometime in the 1980s, in 1997 Polmar said that the latest Soviet nuclear submarines were actually *quieter* than the U.S. improved *Los Angeles* class, at least at tactical speeds of five to seven knots.[47] In any case, the Soviets/Russians, like so many others, have many high-quality photos of American carriers taken by surprise and at close range.

7

The Chinese: Know Thy Potential Enemy

THE CHINESE TOO HAVE A STRONG interest in neutralizing American aircraft carriers. In his 2000 book *China Debates the Future Security Environment*, Michael Pillsbury demonstrates that the Chinese have completed detailed studies of the vulnerabilities of U.S. Navy carriers. He documented that the Chinese have noted the following possible weaknesses: lack of stealth due to the large number of radar reflections plus infrared and electromagnetic signatures, all of which make the carrier "very difficult to effectively conceal";[1] flight restrictions during bad weather; inability to operate safely in shallow waters; decreased readiness during regular at-sea replenishments; poor ASW and mine countermeasures capabilities; and structural vulnerabilities of catapults, elevators, and arresting gear.[2] Sun Tzu put it best when he said the immortal words, "Know thy enemy and know thy self and you will win a hundred battles." It seems the Chinese have taken Sun Tzu's advice to heart when it comes even to their potential rivals.

These days many analysts are quite concerned about a possible confrontation between the United States and China over Taiwan. While O'Hanlon suggests that China does not have the necessary means to invade and occupy Taiwan, others feel the Chinese might still attempt to do so.[3] If that does happen, it would be well not to underestimate the Chinese. One need only recall Appleman's book *Disaster in Korea: The Chinese Confront MacArthur* to see that the low-tech Chinese have been a most dangerous and wily opponent for American forces. In 2004, Goldstein and Murray (the latter is a former U.S. Navy submarine officer) predicted that if the U.S. Navy comes up against Chinese AIP and diesel submarines blockading Taiwan: "We find a plausible worst case would yield nearly fourteen U.S. ships sunk after a single tactical exchange. Playing out this model to its logical conclusion (iterations until all submarines in the People's Liberation Army Navy are destroyed) with these revised inputs suggest that more than forty U.S. Navy (U.S.N) ships could be sunk."[4] Thus, in their view, the Chinese submarine deployment of roughly twenty-four nonnuclear submarines would be destroyed eventually, but they could take as many as forty American ships down with them, and American aircraft carriers would "not be immune from submarine attack," even if they remain in the comparatively low-risk deep waters to the east of Taiwan.[5]

In another startling and plausible hypothetical U.S. Navy versus China knock-down, drag-out brawl over Taiwan, Corpus predicted that after blinding U.S. satellites and attacking with existing Chinese rocket torpedoes and supersonic cruise missiles, "In less than an hour . . . all the aircraft carriers and their escorts of cruisers, battleships and several of the accompanying submarines are in flames, sinking or sunk, turning the East China Sea and the Philippine Sea into a modern-day 'Battle of Cannae.'"[6] Meanwhile, in the same scenario, U.S. carriers in the Persian Gulf were simultaneously attacked by Iranian forces, which launch a prearranged and internationally coordinated

> barrage of supersonic Granit, Moskit, Brahmos and Yakhont cruise missiles carried by trucks or hidden in man-made tunnels all along the mountainous shoreline of Iran fronting the Persian Gulf. The three U.S. aircraft carrier groups that entered the Persian Gulf to ensure the unhindered flow of Arab oil are likely to be helpless 'sitting ducks' against the bottom-rising sea mines and low-flying, supersonic antiship cruise missiles in Iranian hands. In the process, a couple of oil tankers about to exit the Strait of Hormuz are hit with the aid of rocket-propelled sea mines, thus effectively blockading the narrow strait and stopping oil supplies from coming out of the Middle East. A 'weak' nation like China or Iran, without a single aircraft carrier in their respective navies, could thus obliterate the carrier battle groups of a superpower. Here, one can see the hidden and often unnoticed power of asymmetric warfare, which may well spell the end of 'gunboat diplomacy' in the not so distant future.[7]

The main reason for the predicted relatively heavy U.S. losses is the degradation and withering of U.S. ASW, anti–cruise missile, and MCM capabilities since the end of the Cold War. The Chinese have noted all of this and keep it in mind when they plan their exercises. The Chinese, for good reason I should suppose, are growing more and more confident in their ability to tangle with the U.S. Navy. Said one Chinese senior officer in 2002, "The U.S. likes vainglory; if one of their aircraft carriers could be attacked and destroyed, people in the U.S. would begin to complain and quarrel loudly, and the U.S. President would find the going harder and harder."[8] Another way to look at it is as follows:

> China may not possess any of those expensive aircraft carriers of the superpower, but it can wipe out those carrier battle groups with a 'single blow' of its assassin's mace or *shashaujian*—its major tool for conducting asymmetric warfare to defeat the U.S. in a major confrontation over the Taiwan issue or other issues. The U.S. may possess the most powerful war machine in the world, but it can be defeated by an inferior force by avoiding the superpower's strength and exploiting its weaknesses. Again, an integral part of Chinese doctrine is: "Victory through inferiority over superiority." One famous Chinese strategist, Chang Mengxiong, compared asymmetric warfare to "a Chinese boxer with a keen knowledge of vital body points who can bring a stronger opponent to his knees with a minimum of movement."[9]

8

Lax Security

ONE WOULD THINK THAT the U.S. Navy would spare no expense to protect its bases, especially those in which its nuclear submarines, both attack and missile boats, are stationed. One would think that effective, vigilant, round-the-clock, airtight, multitiered security would shroud an installation in which Trident missile submarines are based. One would think that the security around these nuclear missile-launching platforms would be almost impregnable. But if one also thinks that strong security measures were the norm in the U.S. Navy during the Cold War or have been since then, one should think again.

In June 2001, Lt. Cdr. Jack Daly, USN, told the audience of a radio broadcast called *Judicial Watch* that American nuclear submarine and aircraft carrier bases were becoming increasingly vulnerable to attack due to lax security measures. He cited an incident in April 1997 in which a Russian spy ship reportedly used a laser to attack a helicopter in the Strait of Juan de Fuca, near two Navy bases. Daly and his Canadian air force pilot had suffered permanent eye damage in the attack; Daly said that it was now routine for Russian spy ships to go snooping around the U.S. Navy bases at Bremerton and Everett, Washington. He also stated that the spy ship that attacked his helicopter had "come to within 1,000 yards of the nuclear-missile-armed USS *Ohio*."[1] The reason why the Russians had gotten so bold, he argued, was that the U.S. Navy had grown complacent and unconcerned about espionage and security. "With the end of the Cold War," he said, "the U.S. Navy had basically let its guard down."

Lax security was also evident in October 2000, just a few weeks before the terrorist attack on the USS *Cole*, when a news team from WABC-TV New York completed a two-month investigation on security at the naval stations at Norfolk, Virginia; New London, Connecticut; and Naval Weapons Station Earle, New Jersey. In all cases, the members of the news team had absolutely no difficulty gaining access to the bases, were never asked to produce identification, were able to sail a small boat within a few feet of American ships without detection, and in Norfolk, the world's largest naval base (home port for seventy-eight ships), the journalists roamed freely, unnoticed and unchallenged for four hours. They shot video of their incursion, and when Representative Jim Saxton of New Jersey saw the shocking tape, he exclaimed, "What you have shown me is absolutely incredible; it's unbelievable!"[2] It is hard to understate how much damage a team of terrorists could have done to the U.S. Navy if they too could have penetrated the security at Norfolk and attacked the ships concentrated at the world's largest naval base. Although it is certainly not common, according to GlobalSecurity.org, at times there have been up to five nuclear-powered carriers simultaneously in port at Norfolk.

Of course, both these incursions happened after the Cold War ended. However, it must be pointed out that even during the Cold War, security at U.S. Navy bases was often very exiguous. Probably the most qualified man to speak on this issue is a former U.S. Navy senior officer, Capt. Richard Marcinko. Although he was once court-martialed for misuse of government property, it is hard to discredit a man like Marcinko when it comes to military operations. A former SEAL with over three hundred combat missions to his credit, he earned thirty-four citations and medals, "including the Legion of Merit, the Silver Star, and four Bronze Stars with a combat 'V' for Valor."[3] He is now an acknowledged expert on terrorism and is frequently consulted on U.S. national TV news and current affairs programs. In the 1980s, during the watch of CNO Adm. James Watkins, USN, Marcinko and his SEAL Team 6 were assigned to test security at major navy bases, and the results of his simulated terrorist raids were very disturbing. His team infiltrated the New London naval base, where nuclear submarines, including missile boats, are based. Marcinko's team had little difficulty infiltrating the base, and it made a mockery of the base security forces. In his own words: "I rented a small plane, and Horseface flew us under the I-95 bridge, wetting our wheels in the Thames as we swooped low. We buzzed the sub pens. No one waved us off. We rented a boat and flew the Soviet flag on its stern, then chugged past the base while we openly taped video of the subs in their dry docks, capturing classified details of their construction elements. The dry docks were exposed and unprotected—if we'd decided to ram one of the subs, nothing stood in our way."[4]

Marcinko's team did far worse during his visit to New London. His men infiltrated several nuclear submarines and proceeded to wreak havoc therein. "First, they found the sentries—who were secure in their shacks drinking coffee—and silenced them. Then, they concealed explosives behind the diving planes of one nuclear sub. They boarded another Boomer [ballistic-missile] sub and placed demolition charges in the control room, in the nuclear-reactor compartment, and in the torpedo room."[5] They were challenged by base personnel but explained that they were just doing maintenance, and amazingly, they were never asked to identify themselves. Marcinko later briefed a very unhappy admiral and boasted, "I blew up two of your nuclear subs, and if I'd wanted to, I could have blown 'em all up."[6]

Karam also reminisced that the SEALs also clobbered his submarine, the USS *Plunger,* during an annual drill. "One year, they swam across the shipping channel from North Island, 'shot' our topside watch and were in control of Control and Maneuvering within a few minutes."[7] To be fair, the U.S. Navy is now taking security much more seriously, but only as a result of the attack on the USS *Cole* and the September 11 attacks. Despite the lessons taught by Captain Marcinko and his SEALs in the 1980s, little was done to improve security in the interim. Apparently the U.S. Navy prefers to learn its lessons, when it does actually learn, the hard way.

Ironically, the SEALs themselves, as good and tough as they are at testing the security of main force units, are not exactly tight-lipped and religiously committed to good OPSEC (operations security), either. As one U.S. Air Force pararescueman said in 2003, the Navy SEAL often makes himself well known to anyone and everyone who will listen, and he proves it "by getting in a bar fight."[8] Another said one can always differentiate the SEALs from other special operations warriors because they are cockier. They are brainwashed, even more so than other U.S. elite forces, and are told constantly "how good they are, and there are instances where they are good, and there are instances where they are bad."[9] Of course, these criticisms may be somewhat tinged by interservice rivalry, but they still have a point.

It is also notable that, unlike many other U.S., British, and French troops, very few SEALs had any combat experience at all during the 1990s, even during Desert Storm.[10] Of course, this fact was not often mentioned in the numerous "rah-rah-rah" style documentaries and books put out at the time. In more recent actions in Afghanistan, the performance of SEAL units was quite uneven, and the SEALs have drawn much criticism for it.

As Maj. Donald E. Vandergriff, USA (Ret.), put it, "While several SEAL units performed outstandingly, overall the SEALS were found to be wanting."[11] This matter is given due coverage in Sean Naylor's refreshingly critical and brutally honest book *Not a Good Day to Die: The Untold Story of Operation Anaconda*, which I highly recommend.

9

A Few Realistic Men

> Operating against a carrier is too easy. The carrier's ASW protection often resembles Swiss cheese.
>
> —Capt. John L. Byron, USN (Ret.)[1]

> My own experience (in war games) is that I never have any problem getting a carrier. . . . [T]hose fleets are going to get ground into peanut butter in a war.
>
> —U.S. Navy submarine commander [2]

> One enemy diesel submarine lucky enough to get one torpedo hit on a CVN (nuclear powered aircraft carrier) or an AEGIS cruiser could easily turn U.S. resolve and have a huge impact on a conflict. . . . [T]he challenge of finding and destroying a diesel submarine in littoral waters can be nearly impossible. . . . In general . . . a diesel submarine operating on battery power is quieter, slower, and operating more shallow than a nuclear submarine.
>
> —Lt. Cdr. Christopher J. Kelly, USN[3]

EARLIER, I DISCUSSED HOW EASY it is for foreign diesel submarines and air forces to attack American carriers. But it is not just the Russians, Chinese, Canadians, Chileans, Dutch, Australians, and former secretary of defense Dr. James Schlesinger[4] who have reason to think the U.S. Navy's carrier strike groups are oversold, expensive, and extremely vulnerable. U.S.

Navy officers have made such arguments often enough. In the 1930s, a U.S. naval aviator said: "Carriers combine great power with extreme vulnerability."[5] In 1939, another senior U.S. Navy officer remarked, "The vulnerability of our carriers constitutes the Achilles heel of our Fleet strength."[6] These remarks were true then, and they remain true today. It is also well known that the cantankerous Adm. Hyman Rickover, USN (Ret.), did not think much of his own carrier-centered navy. When asked in 1982 about how long the American carriers would survive in an actual war, he curtly replied that they would be finished in approximately forty-eight hours.[7] Former president Jimmy Carter, a former U.S. Navy officer and an Annapolis graduate, was also none too keen on the big-carrier Navy. Vistica mentions that Carter did not want any more new carriers, and did want the existing fleet to be cut dramatically.[8] John Lehman has said that former secretary of defense Harold Brown "made no bones about his belief that the navy was of quite secondary utility."[9]

As Byron noted above, it is not a big challenge for a submarine to attack an aircraft carrier. In a 1985 exercise in the Pacific, this was confirmed when one U.S. nuclear submarine sank two aircraft carriers and eight other ships; per standard operating procedure, these painful results "were never publicly disclosed."[10] Shuger in 1989 noted, "I've seen enough photos of American carriers through periscope crosshairs—most sub crew offices feature one—to become a believer. Despite all the antisubmarine warfare equipment that carrier groups take with them to sea, in my own experience most exercises against subs ended up with my carrier getting a green flare at close quarters, the standard simulation for a successful torpedo or cruise missile attack."[11]

The respected naval affairs analyst Norman Polmar said in 1998, "It's just too easy for a diesel sub with even conventional torpedoes, let alone high-speed advanced torpedoes . . . that the Russians are selling, to get a shot and hit a carrier. . . . That could really cause us problems."[12] Former CIA director Adm. Stansfield Turner, USN (Ret.), has proffered many dire warnings that the U.S. Navy's continuing policy of building and deploying "big, over-powered aircraft carriers" is "ill-advised." In his 2003 article "Is the U.S. Navy Being Marginalized?" Admiral Turner submitted that "the day of large aircraft carriers with large numbers of high-performance aircraft is simply drawing to a close. . . . With more accurate weapons, the ordnance-carrying capacity of the large carrier will no longer be as important. On the defensive side of the technology coin, we must recognize that technologies that make our forces more lethal will be available to others.

When opponents acquire remote sensing and precision, long-range targeting capabilities, as they are bound to do, the huge detection signature of the hundred thousand tons of steel in one of today's aircraft carriers will be a tremendous liability."[13] He also noted: "Our existing carriers will have plenty to do for the remainder of their operating lives, but a Navy built around these ships will not carry us into the emerging era of warfare any better than did the USS *Arizona* into World War II. *To procure more large carriers today and expect them to be useful into mid-century is to be blind to reality*"[14] [emphasis mine].

The late Rear Adm. Eugene Carroll, USN (Ret.), himself a former aircraft carrier skipper, was also an outspoken critic of the Navy, with its infatuation with big aircraft carriers and its collective fear of change. He once said that if the United States continued on its path of building ever larger and ever more expensive aircraft carriers, it would eventually degenerate into a "bankrupt nation." The most damning comment ever made by a senior officer was that of the past Chief of Naval Operations, the late Adm. Elmo Zumwalt, USN, who in 1971 confessed that with the advent of long-range Soviet antiship missiles, if there had been a U.S.-Soviet conventional naval war, the U.S. Navy "would lose."[15] More to the point, according to Marolda and Schneller, "Just before he retired as CNO in 1974, Admiral Zumwalt observed that 'the United States has lost control of the sea lanes to the Soviet Union.' He observed years later that 'the odds are that we would have lost a war with the Soviet Union if we had to fight (during the mid-1970s); the navy dropped to about a 35% probability of victory.'"[16]

Even during the happy days of the Reagan buildup in the 1980s, the U.S. Navy was still not in a great position to fight the Soviets in the distant waters of the western Pacific, because as Hayes et al. noted, even "a 'half' war like Vietnam consumed huge quantities of supplies. In 1968, U.S. forces used 44 million barrels of oil and over 1 million tonnes of ammunition alone. Whether U.S. Pacific logisticians can supply similar volumes over greater distances without collapsing into chaos is doubtful. As late as 1984, CINCPAC [Commander in Chief, Pacific Command] strategists had reportedly not even calculated the basic lift requirements for supply across Pacific Command's vast distances."[17] Indeed, "To sustain a carrier task force at war in the Indian Ocean will be quite difficult, requiring two URGs (underway replenishment groups) resupplied at Subic Bay. Three carrier task groups fighting into the western Pacific and Indian Ocean would require up to nine URGs to support the 7,600 km pipeline," they said. "If,

Adm. Elmo R. Zumwalt Jr. was certain that the U.S. Navy could not beat the Soviet Navy in the early 1970s. —*E. Fredette, U.S. Navy photo*

as proposed by Navy Secretary Lehman, *fifteen* carrier task forces fight a long-distance, global war simultaneously with the Soviet Union, they will wallow helplessly in the ocean since there are simply not enough URGs to supply them."[18] Even now, as plans unfold to build more surface combatants for the U.S. Navy, which I think are long overdue, there are those who say that the Navy is still not building enough support ships to sustain the fleet on its own, and thus dependence on foreign navies will probably continue into the foreseeable future.

If Zumwalt was correct, the only way the U.S. Navy could handle the Soviet Navy would have been nuclear weapons, which in turn would have provoked a Soviet response, after which, in all likelihood, both sides would have been destroyed. Apparently, Adm. Thomas Moorer, USN, had reason to worry also. When Soviet and U.S. ships confronted one another in the

Mediterranean during the October War of 1973, as Goldstein and Zhukov observe, "Soviet battle groups were using the actual U.S. aircraft carriers in the area as virtual targets, an act comparable to holding a cocked pistol to an adversary's temple. Adhering to a kamikaze-like, 'battle of the first salvo' doctrine, the Soviet force of 96 ships was poised to launch approximately 13 surface-to-surface missiles (SSMs) at each task group in the U.S. 6th Fleet deployed in the Mediterranean. U.S. Adm. Elmo Zumwalt, then chief of naval operations, recalled a Washington Special Action Group meeting at the peak of the crisis, during which Admiral Thomas Moorer, chairman of the Joint Chiefs of Staff, estimated: '[W]e would lose our [expletive] in the Eastern Med [if war breaks out].' "[19]

Indeed, decades later, a submariner who was aboard a participating Soviet nuclear submarine stated that U.S. ASW forces were not able to defeat the Soviet forces: "During the events of 1973, our submarine carried out its service for sometime in the vicinity of the Sidra Gulf, by the Libyan coast. Here, a group of U.S. Navy antisubmarine ships, evidently acting on some intelligence, or maybe presuming that there might be a Soviet submarine about, was vigorously carrying out a search operation for two days. However, we gathered the impression that the ships achieved no success. Nothing suggested that our boat had been discovered, even though we were thoroughly listening to their hydroacoustic transmissions and sometimes the hum of the ships' propellers."[20] He also said, "I think that [the Soviet submarine fleet] would have withstood [a U.S. first strike]. . . . There was no reason to believe that our submarine had been discovered by the probable foe . . . in October 1973. If so, then it is entirely possible that we could have been the first to deliver the blow."[21] Given the American predilection for letting the enemy strike first (or, as in Vietnam, to claim the enemy struck first), it is reasonable to assume that the Soviets might well have struck first, and if they had, the American ships in the area would probably have been destroyed or incapacitated very quickly, possibly before they could have fully retaliated.

Another senior American officer who might agree with Zumwalt, Rickover, Turner, Carroll, Byron, and Shuger is Lt. Gen. Paul Van Riper, USMC (Ret.). In Exercise "Millennium Challenge" (2002), Van Riper, playing the role of Saddam Hussein, used small boats to destroy sixteen U.S. Navy ships, including an aircraft carrier and two helicopter carriers, in the Persian Gulf. As usual, the U.S. Navy was not pleased with this successful attack against its most powerful ships, and so it stopped the exercise, "reactivated" the dead ships, and continued as though nothing had happened. "A phrase I heard over and over was, 'That would never have happened,' Van Riper recalls. And I said 'Nobody would have thought that anyone would fly an airliner into the World Trade Centre' . . . but nobody seemed interested."[22] The irrepressible Stan Goff, in his own erudite yet polemical style, explained why this had to be: "The reason Van Riper's victory had to be overruled is that it tears the scary mask off the bully and lets the whole world see the fundamental weakness of the vastly complex and expensive U.S. military monstrosity—the one that will invite not less but more 'asymmetric warfare,' the very monstrosity that is already mortgaging our children's future."[23]

Please note that Goff called the U.S. a "bully," not I. Sadly, this kind of official denial is standard operating procedure in the U.S. Navy. Consider also the American submarine commander who once said that during war games he "put six torpedoes into a carrier, and I was commended—for reducing the carrier's efficiency by 2 percent."[24] The battleship admirals played the same mind and word games when they ran the Navy, and we all know what happened to the battleship.

Many of the criticisms of the carrier-centered Navy come from U.S. Army officers who see the Navy more as a rival than a partner in national defense. One might dismiss such criticisms as merely parochial slander, but some Army critics make good sense. Lt. Col. Douglas Macgregor made a number of convincing arguments in his ground-breaking book *Breaking the Phalanx*. Macgregor is a vocal critic of American military strategy, and his criticisms are not restricted to the Army. He argues that with the U.S. Navy's new focus on littoral warfare, the big-carrier force is in even more danger than during its days as a high-seas fleet designed to face the Soviet Union. The fact that American aircraft carriers are so big, with so much firepower concentrated within them, makes them attractive and worthy targets for weapons of mass destruction in littoral waters: "The concentration of several thousand sailors, airmen, and Marines in an amphibious or *Nimitz*-class aircraft carrier risks single point failure in future warfighting."[25] Also, as the quality and availability of cruise missiles increase, so do the chances of a successful attack on carrier battle groups: "The survivability of large carriers and amphibious ships depends on antiship missile defenses, which must perform perfectly within a few seconds of a missile alert. In both cases, very expensive platforms can be destroyed by relatively inexpensive weapons."[26]

Inexpensive weapons, like antiship missiles fired from speedy hydrofoils, have indeed made life difficult for aircraft carriers in free-play exercises, for example. Recently, Capt. D. Michael Abrashoff, USN, described in detail how his friend, Capt. Al Collins, while serving on the hydrofoil USS *Pegasus,* once clobbered a carrier.

> Lying quietly in wait with all transmitting equipment shut down, the hydrofoil would listen for electronic emissions from the battle group—not just radar signals, but virtually anything in the electromagnetic spectrum, including radio transmissions and the signal waves generated by running engines. Al would triangulate the source and calculate where to aim the missiles. In the simulated firing, a radar ping would signal each of five missiles headed for the carrier in Al's sights. . . . 'One hundred percent of the time, we were successful in getting in, attacking, and then sprinting out of there before they knew what hit them,' Al told me.[27]

Williscroft said in September 2004 that there are several possible nightmare scenarios that face the modern U.S. Navy, and they most certainly involve quiet conventional submarines: "The bad guys can station one of the new ultraquiet AIP subs at a choke point, and seriously damage or even sink a carrier. An AIP sub can sneak up on a *Virginia*-class [nuclear submarine] deploying a Seal team with devastating results. A hunter-killer pack of several AIP subs can take out any nuke we have, once they find it."[28] Macgregor also noted that at a cost of approximately four billion dollars for construction alone, the loss of even one *Nimitz*-class carrier would be morally and financially devastating. The loss of one or more of the two-billion-dollar *Virginia*-class nuclear submarines would also be a tremendous burden on the U.S. Treasury.[29]

10

This Isn't *Top Gun*—and Watch Out for the Little Guy

> U.S.N. pilots worry more about being able to come aboard [i.e., land on carriers] than about their tactics. It is not totally unreasonable, especially in bad weather, night operations. Fortunately, for the U.S.A.F., a landing makes about the same demand as breathing, and frees them to concern themselves with the tactics and doctrines of aerial combat.
>
> —Col. Everest Riccioni, USAF (Ret.)[1]

As we have seen, U.S. carriers are remarkably vulnerable to attacks by submarines and aircraft, but what about the much-vaunted American naval aviators? How would the U.S. Navy pilots fare in air combat maneuvering (ACM) with a much smaller, less powerful, but well-trained enemy? The evidence is not encouraging. Consider Canada, for example. Often criticized by U.S. and NATO officials for very low defense spending (about 1.2 percent of Canada's GDP is spent on defense), Canada fields armed forces that are among the smallest in the alliance (currently about sixty thousand in the regular army, navy, and air force, *combined*). These days, America's northern neighbor, as Charles Moskos observed, comes close to fitting into the "Warless Society" classification in his national military taxonomy, in which military expenditures remain small in peacetime and "the bulk of the military budget consists of personnel training costs."[2] Of course, Canada is now fighting the Taliban in southern Afghanistan, but it is still not a "big spender."

Despite this chronically low peacetime funding (and even when both countries were fighting World Wars I and II and Korea, Canada's military budget was never even close to that of the United States) Canadian pilots have routinely outperformed U.S. Navy and Air Force aircrews in combat and in peacetime exercises. It is easy to understand why if one also comprehends that training, cohesion, and professionalism are the major factors in combat success, neither of which need to be excessively expensive, and the Canadians have a strong record in this area. Here, just for illustration, if you will indulge me for just a moment, are a few surprising highlights related to the Canadian war record, of which most Americans are unaware. In World War I, a group of just ten Canadian fighter pilots was responsible for shooting down a jaw-dropping 438 German aircraft.[3] Even though many Americans sincerely believe that it was latecomers like the dapper Capt. Eddie Rickenbacker, in his stylish, French-designed Spad VIII, that won the Great War in the air, the Canadian, French, and British pilots were the true eagles flying over the western front. Of the top ten allied aces (pilots with five or more air-to-air victories), four were from Canada (the top-scoring Canadian had seventy-two confirmed kills), one was from France (the Frenchman was the top ace), and the others were British. Canada, which at the time had fewer than eight million people (which is roughly the same number that New York City has today), produced 185 aces.[4]

To sum up, says Lt. Col. Martin A. Noel Jr., USAF (Ret.), most American fighter pilots are to this day profoundly unaware of "the significant contribution of Canadian fighter pilots to the history of airpower," which includes being the first to "intercept a German airship over England; the first to spot for artillery at night, using flares," and being only the second to use planes to attack submarines. Canadian fighter pilots were also the first to make the change from single-aircraft fighter tactics to formation tactics utilizing "the 'finger four' formation, known to U.S. airmen down through the next 60 years or so as 'fluid four.'"[5]

The renowned British military historian Sir John Keegan has made reference to the "legendary fighting qualities of the Canadians"[6] in World Wars I and II and has described the Canadian Forces of today as "highly professional."[7] In World War II, Canadian Spitfire pilot Buzz Beurling shot down twenty-seven enemy aircraft (confirmed, plus three other probable kills and eight damaged) *in only fourteen flying days*. He was nicknamed "The Falcon of Malta," and as Lt. Col. Rob Tate, USAFR, explained in 2004,

Beurling's triumphs were "one of World War II's memorable aerial-combat achievements."[8] Further, as Dan McCaffrey noted, in World War II "Canada produced more aces per capita than the Americans,"[9] though one must keep in mind that Canada entered the war two years before the United States did. Lt. Col. James G. Diehl, USA, has also recorded that the contribution of Canadian fighter pilots in World War II "was enormously significant and out of all proportion to [Canada's] population and industrial capacity."[10]

In the 1950s, Canadian pilots flying the Canadair Sabre Mk. VI, with its souped-up Avro Orenda engine, "flew circles" around every fighter plane in NATO.[11] From 1955 to 1958, Col. Everest Riccioni, USAF (Ret.), has said, he "used to instruct my squadron mates in F-100Cs operating against experienced, gutsy, competent Canadian pilots flying Mark VI Sabres." He praised the Canadian pilots as highly skilled and so aggressive that they "would rather fly through you than lose."[12]

During the days of Royal Canadian Navy carrier aviation it was well known that the pocket carrier HMCS *Bonaventure* could at times put more planes in the air than much larger U.S. ASW carriers of the *Essex* class. Said Lieutenant Commander Roland West, RCN (Ret.), in 2005, "Having served in *Bonaventure* on many occasions, I can attest to the fact that there were times when our aircraft flew operational missions when our NATO allies decided to keep their aircraft on the deck. You can rest assured that at no time did our operational commanders put our aircraft and crews in flight safety jeopardy. As far as performance is concerned, there was always that pride in carrying out a role in such a manner that encouraged good competition and success in the operation at hand."[13] This was confirmed by Soward in his description of a 1969 exercise:

> The operations of VS 880 and HS 50 meanwhile continued with the scheduled sustop [sustained operations], in spite of the high wind states and increasingly rough seas. Other NATO units terminated their carrier flying but *Bonaventure* pressed on with ASW operations and continued prosecuting submarine contacts and recording "kills." In keeping with past performance, the carrier logged more flying hours with the Trackers and Sea Kings than any other carrier involved in the exercise.... Unfortunately, about the same time, the U.S.N. carrier USS *Yorktown*, also a major participant in the exercise, suffered the loss of a Sea King and three crew members.[14]

Furthermore, the fighter pilots on *Bonaventure* were also tasked with North American Air Defense (NORAD) continental air defense duties while ashore, and they often beat U.S. Air Force squadrons in the number and quality of interceptions.[15] Also, Lieutenant Gordy Edwards, RCN, who served as an exchange pilot with the U.S. Navy in 1960, said that Canadian pilots were, by default (consider the geography), more experienced in bad-weather flight operations.[16]

This helps explain why Canadian aircraft can continue to fly in rough weather when the U.S. Navy prefers to stand down. Former U.S. naval aviator Thomas Tomlinson felt the same way. Tomlinson joined the Royal Canadian Air Force as a pilot before America entered World War II, later becoming a pilot in the U.S. Marine Corps. He has noted that many of the Americans who joined the RCAF and who then became U.S. naval aviators "had more than a little criticism to offer the Navy" when it came to flight training.[17] In particular, he noted the RCAF-trained pilots were far better trained in instrument flying, which is necessary in poor weather conditions, and that the U.S. Navy's instrument training regimen was "really hilarious" by comparison.[18] And as for ground crew, a Canadian who trained with the U.S. Navy at Key West, Florida, in the early 1960s concluded that the men in the RCN's aviation trades (rates) were "superior in skill to equivalent U.S.N."[19] personnel, due in large part to the U.S. Navy's tendency to overspecialize its personnel (more on this later).

In the early 1980s it was revealed that the average pilot in the Canadian air force flew about three hundred hours a year, whereas his U.S. Navy counterpart flew only about 160 hours annually and U.S. Air Force pilots only averaged about 120 hours (remember, Air Force pilots do not need to fly as many hours, because they do not land on aircraft carriers). This gap might explain why in 1977 a Canadian team flying obsolescent CF-104Gs beat U.S. Air Force F-15s and F-111s to win the NATO Tiger Meet at the RAF base at Greenham Common. Canadian CF-104Gs once again beat U.S. and NATO aircraft to win at the 1979 and 1981 Tiger Meets.[20]

F-104s can beat F-15s? Yes, it is true. I know several Canadian fighter pilots, one of whom told me that he had successfully "bounced and killed" USAF F-15s from the cockpit of an old CF-104G, *and the Starfighter was not even designed for ACM*. As the 1989 Australian documentary "Great Planes: The Lockheed F-104 Starfighter" mentioned, "There is a considerable body of opinion . . . that even suggests that, in aerial combat, the Starfighter itself might give the F-15 trouble."[21] It has indeed, which is not bad for an aircraft that first flew more than fifty years ago! Fighter pilot and author Lieutenant Colonel David Bashow, Canadian Forces, told me

> I personally acquired quite a bit of gun camera film of Eagles and the like over the years. The trick was to fight that class of jets on *your* terms, not *theirs*. On the plus side, the '104 was very tiny and difficult to visually acquire. It was also blindingly fast, especially at low level, with a tremendous acceleration capability. However, it did take just about all the airspace over the Province of Alberta to do a decent turn, so you did not normally enter a turning engagement with just about anything. That said, below 5000 feet, light and in full afterburner, it could *sustain* 7Gs, but that was way up at around 480 knots, *so we are still talking big turns* (and lots of gas pissing out the back end), so you did not do even that for long. The idea was to use the vertical a lot and go for quick kills of opportunity, unobserved, better still. However, you would be surprised how often that would occur, particularly in a multi-bogey environment, and that would *certainly* have been the case in an air war over Central Europe, for which we trained.[22]

By the late 1980s, Canadian fighter pilots were at the top of the charts in NATO, flying more hours per year than all other allied forces in Europe (German pilots came in second and USAF pilots third.) One U.S. Air Force senior officer noted that Canadian CF-18s "performed superbly in the Gulf War,"[23] and in 1996 the famous American pilot and author Col. Walter Boyne, USAF (Ret.), rated the Canadians and Israelis as the two most challenging foes for top U.S. fighter pilots on exercises.[24] That same year, just five years after the U.S. Air Force dominated the air battles in Operation Desert Storm, a Canadian fighter team defeated all comers (six U.S. Air Force and Air National Guard teams) at the prestigious William Tell competition. Some say no team in history had been as dominant as the Canadians were. They won accolades for Top Gun (for the third time), Top Team, Top Operations, Top Element, and Top Weapons Director Team.[25]

A U.S. Air Force three-star general, the Joint Force Air Component Commander, also spoke very highly of Canadian CF-18 fighter capabilities during the Kosovo air campaign in the late 1990s.[26] Actually, during that campaign Bashow noted that even though the USAF was the dominant force, "Such confidence and faith was placed in Canadian aircrew professionalism and expertise that, in this massively American-dominated air campaign and despite interoperability problems due to lack of equipment commonality, Canadians were often selected to lead the strike 'packages.'"[27] As Colonel D. A. Davies of the Canadian Forces has summed up: "We led over half of all of the packages we flew. While the Brits led some, the bulk of the remainder was led by U.S. forces. This is indicative of the high degree of professionalism and excellent training of our pilots."[28] "The CF-18s were extremely effective" over Kosovo, noted Lt. Col. Samuel Walker, USAF.[29]

In 2001, a small Canadian team flying four elderly CF-18As did quite well against U.S. Air Force and Air National Guard F-16C/Ds and F-15Es at the Tiger Meet of the Americas (TMOTA) competition, hosted at Buckley Air Force Base, Colorado. "All three flying days of the exercise were planned for a high degree of realism. The assignment of dedicated tankers to both attacking and defending forces enabled missions in excess of four hours duration, and maximum use was made of the UTTR ground threats—with electronic simulations of long, medium and short range Surface to Air missiles (SAMs)," said Hunter, Jackson, and Greengrass of *Sharpshooter* online magazine.[30] "The flying phase was judged to be a great success, enabling high value dissimilar air combat training between units that would otherwise rarely meet, and tactical flying in a high threat environment." At the end of the competition, they said, "The Canadians maintained their reputation as competition star performers—being voted as TMOTA '01 overall winners. Flight operations, beautiful special markings, and a victory in the 'tug of war' all contributed to their success."[31]

Perhaps this victory was part of the reason why, in 2001, then U.S. secretary of state (and former chairman of the Joint Chiefs) Gen. Colin Powell, USA (Ret.), informed the new U.S. ambassador to Canada that the Canadian Forces, despite their tiny budget and all the bad press they get, are "quite good."[32] (All this, by the way, also tends to deflate Tom Clancy's ill-informed 1995 claim that "no other air force trains as hard" as the USAF or "even begins to match" its professionalism.)[33]

Indeed, since the late 1990s, Canada's new military pilot training center has established a new standard of excellence and is recognized internationally as having the most advanced pilot training regimen in the world. The official Canadian Air Force website makes it clear that Canada's pilot training system is far ahead of the U.S. Navy: "To date, Canada has sold more than a billion dollars in training to pilots from Britain, Italy, Denmark, Singapore and Hungary since the inception of NFTC (NATO

Flying Training in Canada) training in 1999. Using the most advanced and effective integrated pilot training system at the most modern training facilities currently available in the world, Canada has become the benchmark in military pilot training. 'We have the leading edge, most advanced technology for pilot training in the world. It is well ahead of everyone, Britain, the United States, everyone. It is the model for other countries so we are very proud of that,'" said Lieutenant Colonel Brian Houlgate, director of the Canadian Aerospace Training Project.[34]

Canadian fighter pilots, in particular, receive certain training benefits that are simply not readily available to many U.S. Navy aviators most of the year, simply because Canada has huge, underpopulated areas that are ideal for flight training. American naval aviators at bases such as Oceana Naval Air Station (the largest U.S. Navy fighter base on the East Coast) must deal with massive military and civilian air traffic congestion, plus the close proximity of civilian living areas, and thus very limited airspace. As a result, according to journalist Jack Dorsey, their training, particularly at low levels, suffers because of safety and noise concerns.[35] Canadian pilots training at Cold Lake, Alberta, Canada's largest fighter base, face far fewer restrictions due to the base's relative isolation and have access to several ranges that cover an area of 270,272 square miles, slightly larger than the state of Texas. That is one reason why many U.S. Navy pilots covet the opportunity to fly at the Canadian base during the annual Maple Flag air combat exercises. But it is not just the vast airspace that attracts the interest of American pilots. The new Canadian air combat training system now in place at Cold Lake "is the first system of its kind" to integrate a "rangeless" air combat maneuvering instrumentation system "with an electronic warfare system—the Surface Threat Electronic Warfare (STEW) system, which simulates surface-to-air and other ground-to-air threats." "Together, these systems make up the most modern training system in the world today," said Keith Shein of Cubic Corporation in June 2004. "The combination of these two training systems enables pilots to realistically view their performance and tactics on each mission."[36] Better training makes better pilots. Canada has a tradition of excellence in aircrew training, and that is why President Roosevelt once called that country "the Aerodrome of Democracy."[37]

And to wrap up, in 2005, Brig. Gen. Jack Sterling, USA, said, "The Canadian Forces are a world-class, fully capable, professional force, and it will be a privilege to work alongside them."[38] In its heyday, said Col. Austin

Bay, USAR (Ret.), the Canadian military was "one of the world's most able and elite combat organizations," and even today

> Canada's military continues to attract outstanding men and women. I have yet to meet or serve with a Canadian soldier who failed to impress me with his professionalism and discipline. In my experience—in terms of individual, quality personnel—only Australian troops match Canadians on a one-for-one basis. Two years ago, I had the privilege of serving with Australian troops in Iraq. The Aussies are crack. In the mid-1970s, I had the privilege of working with the 4th Canadian Mechanized Brigade Group in then–West Germany. In my opinion, the Canadian brigade was the best brigade in NATO, which probably meant at that moment in time it was the best brigade man-for-man in the world.[39]

Colonel Bay isn't the only foreign officer who has spoken highly of the Canadian Forces. Indeed, as Thomas Axworthy pointed out, "the commander of the British Army of the Rhine said in 1957 that Canada's brigade was 'the best fighting formation in the world.'"[40]

All of this goes to show that the little guys should not be taken lightly and that they should not be underestimated just because they do not spend a lot of money. As Brock once said, "In the U.S. they spend more on their retirement pensions for senior officers than we do in our whole national defense budget."[41] Even though the U.S. defense budget is thirty to thirty-five times greater than Canada's these days, Canadian naval and air units are often better trained, and in some instances better equipped, than U.S. Navy units.

Now, in the interests of full disclosure, I am Canadian, and my great-great uncle was one of the ten Canadian World War I fighter pilots described above. But the foregoing is not idle boasting, ethnocentrism, or a manifestation of the so-called Canadian inferiority complex. As I said before, I have relatives who served in the U.S. Navy (I am one-quarter American), and I certainly have published articles that were quite critical of the Canadian military. (I have indeed thrown a few stones at my own house, but luckily, I am thick skinned and I am no flag-waving partisan. I do not mind at all that my house might have some glass in it.) Years ago, when I was still in grad school, for example, I drafted an article that strongly argued that the Canadian Naval Reserve should be abolished and its budget given to the regular navy for modernization. My article was submitted for publication but rejected because one member of the review board, a retired RCN officer, said my work was simply too "dangerous" to publish. As I see it, the Canada/U.S. military comparison follows the same asymmetries as the classic David-and-Goliath contest, and I have focused largely on Canadian examples because, simply by default, I just know more about the Canadian military than I do about other minor military powers (although, for good measure, I have also included examples from Australia, Chile, Sweden, the Netherlands, and other countries that have modest but professional armed forces). The point I am trying to make is that, contrary to what most Americans believe, and despite Canada's puny numbers and low funding, the Canadians (Davids, if you will) have indeed had great success competing against U.S. Navy (Goliath) pilots. As for the reason why, perhaps part of the answer is, as Col. Kalev Sepp, U.S. Army (Ret.) has conveyed, "Smaller often allows for better in key skills, the meager Canadian defence budget notwithstanding."[42]

Smaller, more parsimonious countries like Canada neither have nor wish to incur the extravagant costs associated with supercarriers, intercontinental ballistic missiles, and the like, which siphon money from the training budget; they prefer to devote more money and time to training than do the American forces. Nor can Canada and most other nations afford to pay $640 for a military toilet seat, let alone $435 for a claw hammer, which the United States did in the early 1980s.[43] And, contrary to what the mandarins in Washington would have one believe, this gouging by rapacious American defense contractors most emphatically did not end in the 1980s, despite the passing of the 1984 Competition in Contracting Act. In 2000, thanks to sloppy fiscal controls, the Pentagon paid seventy-six dollars apiece for screws that actually cost only fifty-seven cents.[44] In 2003, the *San Francisco Chronicle* reported that the Pentagon could not account for $1,000,000,000,000 (yes, that is one *trillion* dollars) of the taxpayers' money.[45] Perhaps this can help explain why the U.S. Navy has so much trouble finding the money for training, despite having a much bigger budget than any other navy. Unsurprisingly, Mark Zepezauer has noted the only way the Pentagon could have accomplished this would have been through "world-class incompetence or unparalleled deceitfulness (or both)." "When it comes to throwing money away," he said, "The Pentagon has no peer."[46]

Like the Canadians, the Israeli air force, quite probably the best trained and most experienced in the world, has outshined the U.S. Navy, and it has done so more than once. A joint USN-IAF air combat exercise in 1999 underlines and highlights the thesis that the U.S. Navy is overrated. On September 14, 1999, the *Jerusalem Post* announced that the Israelis had soundly dispatched the air wing from the USS *Theodore Roosevelt* (which, incidentally, was the same carrier the Dutch destroyed in 1999). Israeli F-16s squared off against American F-14s and F-18s. The final results were astonishing. The Israelis shot down a whopping 220 U.S. aircraft while losing only twenty themselves. The eleven-to-one kill ratio was so embarrassing that the results were not "officially published 'to save the reputations of the U.S. Navy pilots.'" The magazine article on which the article was based, however, reported the kill ratio to be about twenty to one.[47]

Some dispute these figures and claim that the Israelis had an "unfair advantage," not having included American victories from "stand-off missile hits." (As one fighter pilot told me recently, thanks to their large egos, fighter pilots tend to be quite expert at "cobbling together excuses.") Responding to claims by a U.S. Navy spokesman that the aforementioned victory by Israel was meaningless, former F-14 radar intercept officer Jerry Burns retorted, "He gets paid to say that."[48] And as the *Washington Times* reported on September 15, 2000, the Navy's inspector general, Vice Adm. Lee F. Gunn, USN, was unimpressed with the Navy's performance and its banal excuses. "Navy pilots were thoroughly beaten in an exercise against Israeli fliers. 'An air wing commander was proud the Israelis only achieved a 6-to-1 kill ratio during simulated air-to-air combat maneuvers against a carrier air wing during a recent exercise, instead of the 20-to-1 kill ratio initially claimed.'"[49] Regardless of the exercise parameters and conditions, it is likely that Israeli F-16s would have had the upper hand against U.S. Navy F-14s and F-18s, because the Israeli aviators were much more experienced in ACM and the F-16 is simply a more maneuverable and agile aircraft. As one U.S. Navy Top Gun instructor put it, "Israel has no average pilots. The intimidation factor teaches them to be competitive and to think for themselves, to always rely on their own skills and courage."[50] F-16 pilots often say that the F-18 is a worthy opponent, with its excellent AOA (angle of attack) capabilities, but it is also "a little underpowered with its smaller engines, and decelerates quickly."[51] And as we will see below, the now-retired F-14 "Tom Turkey" (especially the original A model) was never particularly good at ACM, and the restrictions on using long range missiles were probably not very significant, either.

Pilot skill and surprise are key, and as one of my Canadian fighter pilot friends told me, even a "clapped out" 1950s-vintage T-33 trainer can take out F-18s, "depending on the day, aircrew and circumstances."

Although it is preferable to have both superior training and a superior platform/weapons, Chuck Yeager did have a point when he said that "a more experienced pilot will always get the better of you, no matter what you're flying. It's dead simple."[52] Nichols and Tillman made the same point, but with more color: "The moral is obvious: a good driver in a clunker can beat a Gomer Pyle in a hotrod."[53] A great driver and a great machine make the most lethal combination, but if one cannot have both, the former is probably the better choice most of the time.

This incident was not the first time the U.S. Navy had found itself running behind the Israelis in air combat. In 1983, significant qualitative differences between the Israeli air force and American naval aviation became obvious when the U.S. Navy botched a raid over Lebanon to suppress Syrian forces there. Aircrews from the USS *John F. Kennedy* were not properly briefed, launched with the wrong weapons, used outdated tactics, lost 20 percent of their aircraft, and in return, did very little damage to the Syrian positions. The Israelis, conversely, had enjoyed great success during hundreds of missions over the Bekaa Valley with negligible losses. Yes, the Israelis had far more experience flying over the region, and thus a major advantage, but even Secretary Lehman, himself a Naval Reserve aviator, granted that the Israelis were simply more organized, more creative, and had far better planning and tactics than the Americans did. "Their loss rate is much lower because they plan. They don't do things on the spur of the moment. They have preplanning. . . . And they use imagination. They're damn good," Lehman acknowledged.[54] As Wilson declared, in very colorful terms, U.S. Navy pilots were shocked and mortified by their poor showing over Lebanon, especially when compared to the almost immaculate performance demonstrated time and again by the Israelis. One pilot on the *Kennedy* indicated his disgust with the Navy's execrable performance by shouting, "What a fucked-up mission!"[55] Another confided to Wilson: "If the American people ever find out that we sent ten airplanes over there from this carrier to do what one plane could do, they'll never forgive us. I'm embarrassed. . . . I wonder if we learned anything at all from Vietnam."[56]

Lehman mentioned a lack of planning, which often reflects poor intelligence gathering and/or dissemination. Shuger has said the U.S. Navy is notorious for this. The Navy, he has remarked, really has no interest in or respect for the role of its own intelligence community; "Navy pilots launch from carriers on nationally significant missions without learning much of the relevant intelligence available to them."[57] This negligence goes right to the top of the Navy pyramid, in Shuger's view, who illustrated this conclusion with the following: "One example of how this blasé attitude runs rampant even at high levels is the 'Admiral's brief' at the Miramar air base where I was stationed when not overseas. Although we squadron intel officers spent considerable time preparing and presenting these weekly briefs, over the three and a half years I attended and/or gave them, neither the fighter wing Admiral nor his Chief of Staff *ever* showed up. In fact, in all my time at Miramar, I only spoke to an Admiral once. That was when I was more or less ordered to participate in a skit for his farewell party at the 'O' Club."[58]

Curiously, that same year (1983) the U.S. General Accounting Office revealed that the U.S. Navy had consistently exaggerated its aircraft "mission capable rates."[59] The GAO said, "Current guidance allows aircraft to be reported mission capable although they—cannot perform the primary warfare roles for which they were designed and procured, and—have been designed for certain systems the Navy deems mission essential, but are missing the systems. It is possible therefore for an F-14 fighter aircraft, for example, to be rated mission capable even though it cannot launch air-to-air missiles, or if it is missing an APX-76 identification friend or foe interrogation set."[60] When a fighting organization uses such loose and meaningless readiness indicators, it is bound for trouble.

The Israelis, on the obverse, have no need to exaggerate their readiness. Aviation expert Lon Nordeen came to the conclusion that "the I.A.F. pilot-training pilot program was the most rigorous in the world" in the 1950s, and by the mid-1960s, the IAF had ultimately developed "air combat skills unmatched anywhere in the world."[61] Both of these assertions were substantiated clearly and dramatically in its combat record in the 1960s and 1970s. As Shlomo Aloni recorded in 2004, Israeli fighter pilots flying the French-made Mirage IIIC and the Nesher (made in Israel but very much a Mirage V type aircraft) and badly outnumbered, scored 397.5 air-to-air kills between 1966 and 1974.[62] "Compared with the U.S. air combat experience in Vietnam, the Israeli aerial kill exchange rate and overall air-to-air performance was phenomenal."[63] Aloni commented that the French Mirage IIIC was technically inferior to its contemporaries serving in the U.S. Navy, but luckily, it did have a cannon, and in the hands of skilled Israeli pilots (who knew how to use obsolete weapons like cannon to great effect) it became the most famous fighter of its era.[64]

Aloni's statement, while impressive and sincere, needs some qualification. If we look at the performance of the U.S. Navy and Air Force's primary fighter of the 1960s, the F-4 Phantom, against Soviet-designed

North Vietnamese fighters in the early years of the war, the overall victory/loss ratio in aerial combat was only about two to one in favor of the Americans. Meanwhile, between July 1966 and the end of 1969, Israeli Mirage and Nesher pilots shot down 116 Soviet-built MiG fighters (mostly with cannon fire) while losing only nine to enemy fighters (a ratio of almost thirteen to one).[65] Many of the enemy fighters faced by both the Israelis and the U.S. Navy were the MiG-21 Fishbed aircraft. The Israelis, as you might expect, had little difficulty handling this opponent. According to Cockburn: "The Israeli Air Force consistently outclassed the Fishbed in the Middle East wars in 1967 up to the engagement with the Syrians over the Bekaa Valley in 1982, destroying an average of 20 for every Israeli plane lost."[66] The U.S. Navy pilots also handled the Fishbed over Vietnam, but not quite as easily. As Rendall remarks, the U.S. Navy and Air Force pilots had far more restrictive "rules of engagement" than the Israelis did, and this no doubt undermined their performance.[67] While these engagement restrictions were certainly not the fault of the pilots, they nevertheless were a key weakness for the U.S. Navy, the U.S. military, and other Western nations who try to fight wars with lawyers as well as warriors. Adversaries who are less concerned with such legal niceties will prove to be quite challenging.

Hallion also cautioned that surface-to-air missiles (SAMs) were much more prolific in Vietnam than in the Middle East in 1967, and that complicated matters for the Americans as well.[68] Others have noted that the Israelis, ironically, had an easier time shooting down enemy planes because there were so many available targets; Vietnam, to the contrary, was not a "target rich" environment for the Americans. Some have noted that the Arab pilots flying against Israel were not of the highest caliber, either. However, there were other major purely self-imposed obstacles that the Americans faced, such as poor ACM skills and using missiles when guns would have been more appropriate (even the most advanced version of the F-4 used by the Navy, the J model, did not have an integral cannon).[69] Spector grieved that "during the spring of 1968, F-4s from the carriers *America* and *Enterprise* fired a total of twenty-seven of these $150,000 [Sparrow] missiles without obtaining a hit."[70] Also, Cockburn also ruminated that the U.S. overreliance on missiles in Vietnam sometimes verged on the tragicomic: "On one occasion during the Southeast Asia war, a U.S. F-4 pilot fired a Sparrow missile at an Australian destroyer because his look-down radar had informed him it was a low-flying helicopter. Luckily for the seaman in whose bunk the Sparrow ended up, the fuze was not working and the missile failed to explode."[71]

The U.S. Navy put too much faith in missiles and technology, but nowhere near enough in training, said Spector. "As one senior officer observed, 'The point is, we sent our people out there not trained for dog fighting. We sent aircraft out there not equipped for dog fighting . . . and occasionally (I probably should use the word frequently) we got into a nose-to-nose combat situation where neither the guy flying the airplane nor the airplane had ever fired a missile."[72] Unnecessarily high losses were the end product of all this, and some U.S. Navy carrier squadrons were badly mauled by North Vietnamese SAMs, AAA (antiaircraft artillery), and aircraft: "Between July and October 1967 the carrier *Oriskany* lost nearly 40% of her combat aircraft. One squadron 'went out with 14 brand new A-4s and something like 12 were shot down. Of course we picked up replacements, but really only one or two of the original planes were with us when we came home.'"[73] The *Oriskany* was not the only carrier to absorb big losses. "In the last year of the war," notes Vistica, "*Saratoga* lost seventeen of its planes. Between 1963 and August 15, 1973, when the war ended, 859 Navy aircraft were lost in combat or operational accidents."[74]

As Capt. Dan Pedersen, USN (Ret.), has said, by late 1967 morale had bottomed out on the USS *Enterprise*. "I came back from that cruise and we had lost seventeen guys [from the air wing] . . . thirteen of them in the last two months of the cruise. . . . If you'd have collectively looked at the air wing crews in those days, there wasn't any happiness. It was dogshit hard work *and we were losing every day*. . . . I lost two close friends in a week. . . . Damn, it was miserable"[75] (emphasis mine). Another pilot recalled that "a lot of squadrons had trouble. People did not want to go into combat."[76] While SAMs and AAA were the main threat, some of these losses were at the hands of North Vietnamese pilots, who "were virtual newcomers to dogfighting,"[77] flying in an outfit that American analysts had previously derided as a "'peasant' air force."[78]

Nichols and Tillman note that toward the end of the war some U.S. naval aviators found ways to avoid combat, which is ironic, since the Americans themselves harshly criticized the South Vietnamese for doing exactly the same thing. They write, "Some pilots reported mysterious glitches that were never diagnosed by the plane captains and mechanics. Radio failure was another way out. All it took was a thumb on the microphone button immediately after turning on the set. Without time to warm up, keying the mike would pop the circuit breaker and the bird was 'down' for that hop."[79]

A former ace in the North Vietnamese air force, Luu Huy Chao, confirmed these U.S. assertions, saying that some demoralized American pilots actually seemed to make themselves scarce when they made contact with his simple but rugged old MiG-17. "Now that the war is over, I want to be completely honest. I only had three hundred .47-caliber bullets. In just a few seconds of air combat I would run out of bullets. But most of the time, when American pilots saw us they flew away in a flash. . . . When I discovered American planes they were always above me and I rose to meet them. Then poof, they disappeared in an instant,"[80] he remarked. "Once I shot down an F-4, I was no longer scared of any type of American aircraft."[81] By the end of the war, Chao was credited with bringing down six enemy aircraft, more than any U.S. Navy pilot achieved.

Chao's MiG-17 was not gold plated by any stretch of the imagination, nor was it as expensive as U.S. fighters, but even modern-day American pilots who have flown this old plane have marveled at its performance. "The 'banana-winged' MiG-17 was 'the greatest turning airplane any of us had ever seen,'" said one U.S. "Red Eagle" pilot in 1997.[82] "It just dazzled us. . . . It was like a boomerang. . . . Shaking a MiG-17 off your six [o'clock position, directly behind] was like shaking gum off your shoe. It made an A-4 laughable."[83] Perhaps it is safe to assume that the U.S. pilots actually had good reason to avoid scrapping with Chao's MiG-17.

Stevenson further complained that the Phantom also had smoky exhaust, making it an easy target; in addition, the canopy restricted the crew's field of vision, giving it a very large blind spot.[84] Plus, even Top Gun graduate Robert L. "Hoot" Gibson, who had flown both the Navy F-4 (in combat) *and* the MiG-21, said that, given a choice, he actually would have preferred flying the MiG-21 in daytime visual flight conditions.[85] It has also been observed that at slow speeds a well-handled MiG-21 can perform maneuvers that the much more expensive F-14 Tomcat could not match.[86] Ironically, if the U.S. forces had used an older fighter, the Canadair Sabre Mark VI, a more powerful version of the standard USAF Sabre as used in Korea, along with better ACM training, they actually might have gotten better results in air combat over Vietnam. The Sabre started out as an excellent all-gun fighter, although later versions were equipped with missiles, and its canopy gave the pilot a 360-degree view. Its engines were smokeless, as well. In the late 1950s, Sabre pilots from the

California Air National Guard, some of them aces from the Korean War, clobbered Navy pilots flying much newer aircraft like the F-8 Crusader and the F-4D Skyray in air-to-air combat exercises,[87] and the Pakistani air force's elderly Sabres had little trouble in defeating Indian MiG-21s by a whopping six-to-one ratio in the 1971 Indo-Pakistani War[88] (although one would have to wonder how the Sabre would have fared against North Vietnamese SAMs).

Acknowledging its poor performance in air combat over Vietnam, in March 1969 the U.S. Navy opened its famed "Post-graduate Course in Fighter Weapons, Tactics and Doctrine," better known as the "Top Gun" course. Here, happily, is an instance in which the U.S. Navy understood it had a weakness and actually did something about it. However, the move was not without controversy. Wilcox wrote that there was great opposition to the new course, and from Navy fighter pilots themselves. "Although the Ault study justified the school, there were still powerful officers opposed to it,"[89] he said, implying that the massive loss of life, aircraft, and prestige over Vietnam was somehow less damaging than possibly having to divert resources from other pet projects or exposing the poor air combat skills of certain senior F-4 naval aviators to ridicule.

Moreover, it can be argued that the Top Gun course, offered to only a very few select crews, who were supposed to refresh their squadrons with what they had forgotten (or never learned) about the art of ACM, was only necessary because the U.S. Navy had not fully trained (or equipped) its pilots in the essentials of close air combat. In effect, at least in the beginning, it might be considered a remedial course (not the "PhD" in air combat tactics) that taught U.S. Navy pilots many of the tactics that the Israelis apparently had learned years before, even though both countries had been fighting air wars since 1966.

The impact of Top Gun, within the limits of Vietnam, is frequently overstated. The ratio of twelve to one that some pundits offer was achieved largely by Top Gun graduates collectively, a very small number of men.[90] Actually, the skills of the *typical* Navy fighter pilot did not change very much. Shuger adds, "Schools like Top Gun are considered such plums that entrance to them is generally restricted to the career-committed 'top one percenters.'"[91] And that it is not really necessary for promotions, because "as long as an officer shows minimal competence in his specialty, he will probably continue to move in step with his peers for a long time, even if his overall military development is totally arrested. All he has to do is come to work and get more senior."[92] "So," he chided, "the great majority of Fleet Average Aviators have been given no good reasons to care about the real nuts and bolts of the next war. For them, flying is prestigious and fun. Isn't that enough?"[93] He also says, "Working on your landing or other airmanship skills shows you have the Right Stuff. Working on your knowledge of enemy tactics and weapons doesn't. That's why most fleet pilots know less about what they'd face in combat than they do about suburban real estate prices."[94]

Even if we discard the rules of engagement and SAM issues, Navy pilots in general lacked the training and weapons necessary to excel in close combat and thus probably would not have achieved the same exchange ratios as the Israelis. And despite launching Top Gun, the top ace of the war was nevertheless a North Vietnamese pilot, Nguyen Van Coc, with nine air-to-air kills (including several F-4 Phantoms). Altogether, North Vietnam produced sixteen aces,[95] whereas the U.S. Navy produced only one ace team (pilot Lt. Randy "Duke" Cunningham and his radar intercept officer, Lt. [j.g.] Willie Driscoll), who together were credited with five victories. The Vietnamese pilots had a certain advantage in that they stayed with their units and engaged in combat for longer periods of time than U.S. Navy pilots. As we will see later, the relatively fast rotation of American pilots from unit to unit has been a serious problem. Cunningham and Driscoll earned distinction by shooting down three MiGs in one day, but in turn they were shot down that same day by a North Vietnamese SAM. Not exactly a Hollywood story, but they were the best the U.S. Navy produced, and Cunningham was a Top Gun graduate.

All this makes the U.S. Navy look pretty bad in comparison to the Israelis, but is it an invidious comparison? Some say that the two organizations and the conflicts they have engaged in are apples and oranges, and perhaps they are correct, at least to a certain extent. This does not let the U.S. Navy off the hook or allow it to exculpate itself, though, because it is also fair to say that the Israeli air combat record is second to none in the postwar era, whereas the U.S. Navy's record is mixed. Lehman was not just being kind when he said that the Israeli air force is "damn good."[96] In the 1973 Yom Kippur War, "the Israeli Air Force had the fastest turnaround time from takeoff to takeoff of any nation on earth,"[97] declared American author Rodger Claire. Analogously, the Israeli student pilots who converted to the F-16 in the late 1970s astounded their U.S. mentors with their in-depth technical knowledge of the aircraft and often knew more about what was to them the brand-new F-16 than the experienced American instructor pilots did.[98]

To summarize, most analysts would concur that the Israelis have been highly successful. Contrarily, a goodly number of analysts would also agree that U.S. naval aviation has suffered many setbacks and humiliations over the years and has had to open special schools to rectify its self-admitted deficiencies. Finally, it would be very difficult to accept that American naval aviators were or are better trained than their Israeli peers, especially in 1983, and considering the U.S. Navy's combat experience over Vietnam, this in itself should be a disgrace. The core issue here, or so it seems, is an apparent inability or unwillingness to learn from mistakes in the long run. Lieutenant Dave Draz, USN, served as an exchange pilot in the Royal Canadian Navy in the 1950s, and he cuts to the heart of the matter:

> I soon learned that the R.C.N. worked just as long and as hard hours as the U.S.N, and just as skillfully. But more importantly, I became ware that "lessons learned" were a part of every day life in the R.C.N. Quite frankly, while the U.S.N was speaking around the clock, the next op[erations] order was being written by staff types with little or no attention being paid to the trials, tribulations and errors being made as events went on. Hence the U.S.N was committed to making or repeating errors without taking any time to say "what if we did it this way."[99]

Now let us briefly compare this somewhat negligent U.S. approach to learning lessons with that of the Israelis. As Cockburn has alluded, during the Six Day War Israel lost one of its largest ships, a destroyer, to Egyptian antiship missiles. The Israeli navy learned its lesson and promptly implemented a solution "by scrapping their few large destroyers and rebuilding their navy around small, fast, and highly maneuverable patrol boats."[100] Although the Israeli Navy is not in any way comparable to the U.S. Navy in terms of missions, responsibilities, or strategic purpose, this example does show that one navy is capable of intelligent and rapid change, even if it means sacrificing its greatest symbols of naval might. The Israelis do indeed learn from their mistakes. One cannot always say the same for the U.S. Navy.

Unlike Israel, Chile is not a great military powerhouse, but its air force is well trained, and it too has given the U.S. Navy reason for pause. In the August 1989 issue of *Air Combat* magazine, Jeffrey Ethell reported that Chilean air force pilots, flying the relatively unsophisticated but nimble F-5E, had trounced a haughty, confident American carrier air group (including F-14s and F-18s) from the USS *Independence* in air combat exercises. The initial kill ratio was reported as fifty-six to sixteen in favor of the Chileans, later revised to thirty-six to twenty, and as one might expect, this incident did not receive much press coverage in the United States.[101]

It should be noted that in this exercise, not unlike the one with the Israelis previously discussed, pilots were not allowed to engage targets that were beyond visual range (BVR), which obviously restricted the use of the U.S. Navy's long-range Phoenix and Sparrow missiles. In addition, aircraft were not permitted to engage in "head-on" or "face to face" missile attacks, which meant that pilots had to maneuver into missile firing position *behind* the enemy.[102] Now this raises an interesting question, namely, were these restrictions unrealistic? Some will say the answer is "not really," and for several reasons. Firstly, ROE have often required pilots to obtain visual identification (VID). For example, U.S. fighters in Vietnam and, more recently those in Operation Provide Comfort over Iraq's no-fly zone (1994), had to identify visually possible targets before engaging them. In the latter case, VID was deemed necessary because "'the No Fly Zone has too many multinational players for aggressive combat.' These peacetime ROE are in place for the sole purpose of slowing down any military confrontation, thus preventing a 'friendly fire' situation that was all too common during Operation Desert Storm."[103]

Secondly, even when the ROE and the combat environment permit BVR combat, as was the case in the atypically one-sided Operation Desert Storm, most of the U.S. air combat kills were still achieved within visual range, out of concern about fratricide.[104] This concern about blue-on-blue disasters is well justified, because, as Riccioni mentioned, no one has yet come up with a truly reliable way to differentiate between friends, foes, and neutral aircraft with anything except the Mark I Eyeball.[105] What Spinney wrote in 1985 on BVR remains true today:

> Air-to-air combat models generally assume that visual identification is *not* required, either because theaterwide rules of engagement permit shooting at unidentified targets or because of a highly reliable identification friend-or-foe system (IFF). Yet a highly reliable IFF system has defied development for years and is still only a projected capability. (If one assumes a highly reliable, thoroughly debugged IFF system, the

combat model can provide simulation outcomes that justify the increasing capability of complex systems. In off-the-record discussions with the editor, middle-level and junior-grade officers and senior enlisted personnel who will actually operate some of these complex systems revealed another set of assumptions. They assumed that most of the complex, computer-driven equipment would either break down or be destroyed or rendered inoperable early in any major war, and that the war would probably have to be fought with maps and grease pencils, messengers on motorcycles, and so on.)[106]

Twenty years later Riccioni confirmed, "This technology has been under development for 35 years, frequently claimed, but never achieved."[107] A diligent pilot with excellent eyesight and exceptional situational awareness is still better, safer, and more reliable than any electronic system, be it IFF or Non-Cooperative Target Recognition. For this reason, "Air Force policy is 'Eyeballs beat machines.'"[108]

This being the case, firing at targets that are BVR, especially with a $1.9 million Phoenix or the $386,000 AMRAAM, especially in crowded airspace that could include commercial (such as an Airbus A300 of Iran Air), friendly, or neutral traffic, is likely to remain risky, certainly expensive, and therefore often to be discouraged in real life and in exercises. Some might also add that after the USS *Vincennes* disaster, in which a U.S. Navy missile destroyed a BVR Iranian airliner, the U.S. Navy is probably the last organization the American public should ever trust when it comes to combat over the horizon. As for the restriction on using missiles from ahead of the enemy, once again Riccioni points out, "Good Navy pilots, smart pilots, prefer to remain away from the enemy's lethal forward quarter,"[109] and the vast majority of air-to-air kills in modern combat have been secured by attacking the rear quarter anyway, so this too was probably of no great salience. Keep in mind also that many countries, including potential enemies, now have all-aspect heat-seeking missiles that can attack a target head on, as well as BVR missiles, not just the United States. Ethell has reported that the Chileans got the jump on the F-14s and F-18s by listening for (and hearing from quite a distance) U.S. radar emissions while simultaneously keeping their own radars on "standby" and flying low to avoid detection themselves. In this scenario, the Chileans attacked from the rear quarter anyway, as the exercise called for an ambush, not a joust, so the restrictions on the usage of the Sparrow and the Phoenix mentioned earlier would, in all likelihood, be mostly irrelevant. Victory in air combat frequently depends on the element of surprise, which the Chileans definitely had and ruthlessly exploited in this scenario.

Surprise, said Fallows, remains the key to victory in air combat: "The most lethal 'ace' of all time, the German flier Eric Hartmann, did everything he could to avoid prolonged 'dogfight' engagements. He claimed that of the 352 planes he destroyed during World War II, fully 80 percent were 'kills' by surprise. On the allied side, one air commander filed a report in 1944 that might as well have been taken from accounts of Korea or Vietnam: '90 percent of all fighters shot down never saw the guy who hit them.'"[110] The problem with BVR, Pierre Sprey indicated, "is that the other plane is also looking for you, and these same radar systems serve as giant beacons, alerting any other plane in the region to your presence. If other planes are equipped with a 'radar warning receiver' [that is, a 'fuzzbuster'], they are quickly aware that someone is beaming radar toward them, and from what direction. The price of powerful radar, then, is to sacrifice off the top the element of surprise that determines 80 percent of all results. And for what? In Vietnam as in all other recent wars, the great majority of 'kills' was not based on radar detection but on the pilot's own visual observations."[111]

Even the new and improved radar-guided missiles available today, which, unlike their predecessors do not rely on the aircraft's high-powered radar to find their targets, can be defeated, because as Lt. Col. Pierre Rochefort, a retired Canadian Forces officer, told me in 2006, "Newer radar warning receivers (RWRs) enable the targeted aircraft aircrew to be warned when a fire-and-forget missile—that's what it's called—is approaching. That stated, the warning time is reduced significantly 'cause the continuous wave (CW) illuminator of the missile puts out less RF energy than what is generated by an airborne fire control radar. . . . That is also why the new defensive electronic warfare systems (RWR, noise and deception jammers, chaff/flare dispensers, and decoys) have automatic features which will initiate a defensive response the moment a CW signal is detected."[112]

The Chileans quickly took down two F-18s, and they were very surprised at how easy it was to detect, stalk, evade, and kill F-14s in particular. (The F-14 is a much larger aircraft than the F-5, and thus easier to find.) Ethell also noted that the Chilean pilots ranged from very experienced to relative novices, so they were not an elite unit, yet he described them as being as good as or better than the intrepid and highly trained U.S. Air Force "aggressor" squadrons.[113] This outcome tends to support Nordeen's

The F-14 Tomcat was a complex and unreliable aircraft and a huge maintenance burden. —*R. O'Connor, U.S. Navy photo*

statement of 2004 that "it has been demonstrated during air wars of the past 50 years that skill, determination, and effective battle planning and tactics have allowed pilots of an outnumbered force of inferior aircraft to overcome the odds and emerge from an air battle—if not an air campaign—as the winner."[114]

The Israelis in their F-16s and the Chileans in their F-5Es have indeed done very well indeed against the U.S. Navy, but do keep in mind that the latter's own adversary pilots, who concentrate much more on air combat tactics than carrier landings and are thus far better than most naval aviators, flying similar aircraft (F-16A/B/Ns and F-5Es and ancient A-4 Skyhawks), have also routinely mopped the deck with Navy F-14s and F-18s. (For a detailed example of how an elderly A-4 flown by an expert can trash an F-14 flown by a novice, consult Robert Wilcox's book *Wings of Fury*.)[115] That the Israelis and Chileans, and others no doubt, could do the same is reasonable, highly likely, and predictable, especially when we recall that by 1995 most of the U.S. Navy's adversary squadrons had ceased operations due to budget cuts.[116]

The F-14 was one of the main victims in these exercises, and the results would not have surprised the legendary Col. John Boyd, USAF (Ret.), either. The F-14 was so big, in fact, that it was sometimes called the "aluminum cloud." To be more specific, in his 2002 biography, Coram states,

> Hollywood and the movie Top Gun notwithstanding, the F-14 Tomcat is a lumbering, poor-performing, aerial truck. It weighs about fifty-four thousand pounds. Add on external fuel tanks and missiles and the weight is about seventy thousand pounds. It is what fighter pilots call a "grape": squeeze it in a couple of hard turns and all the energy oozes out. That energy cannot be quickly regained, and the aircraft becomes an easy target. Navy admirals strongly discourage simulated battles between the F-14 and the latest Air Force fighters. But those engagements occasionally take place. And when they do, given pilots of equal ability, the F-14 always loses.[117]

In 2003, for instance, Robert DeStasio confirmed that Air Force F-15s took on Navy F-14s in a series of three "two versus two" engagements, in which all the F-14s were targeted and no F-15s were lost.[118]

Even the old CF-104G, with a well trained pilot at the controls, could bring down Tomcats from time to time. According to David Bashow: "As far as relative merits of 'others' versus the Tomcat are concerned, all I have to measure from personal experience is '104 vs. F-14, and that was not much of a contest. We were small and if we got in unobserved (or even a late pickup by them), we could do some good work. . . . Yes, the Tomcat is a bit of a truck, and cannot really mix it up on a continuous par with F-16s or late-model Eagles or Hornets in a 'dogfight-type' of engagement."[119] With a good pilot, the MiG-31 Foxhound, which was not designed for ACM, can also successfully tangle with the F-14, by using "hit and run" tactics that take advantage of the Foxhound's superior speed and greater maximum interception range. Yefim Gordon notes, "The American experts' confidence in the F-14A's adequate thrust and acceleration characteristics lasted until the autumn of 1983—and was shaken when the first operational MiG-31s made their appearance in the Soviet Far East. Even though the Foxhound clearly was not designed as a dogfighter, during their frequent encounters with U.S. Navy fighters during patrol missions the Soviet crews would sometimes demonstrate the MiG-31's ability to outrun the Tomcat to their U.S. colleagues."[120] Some Russian sources also state that the "combat potential of the F-14A and its weapons system equals 60 to 70% of the MiG-31."[121] One U.S. official in the mid-1980s opined that "the MiG-31 was superior to any U.S. fighter . . . and had more capable avionics, including a better GCI [ground control intercept] guidance, control and communication system, better air-to-air missiles, and possessed greater speed and a longer combat radius."[122]

The F-14 was to the U.S. Navy what the Sea Harrier was to the Royal Navy; the frontline fighter charged with the defense of the fleet, and on paper, at least, the F-14 appeared to be far superior. In reality, however, the now retired Sea Harrier was both less expensive and much better in ACM. Not only that, but Sea Harriers, based on small British carriers, proved that they could fly in bad weather conditions that had forced American supercarriers to cancel flight operations.[123] According to Commander N. D. "Sharkey" Ward, RN, in the early 1980s Sea Harriers (with American missiles) beat F-14s in mock combat by a ratio of approximately twenty-five to one.[124] The Sea Harrier also totally dominated the skies in the 1982 Falklands campaign, shooting down some twenty-two Argentine planes, with no losses in air combat. Ward also said that Sea Harrier pilots from 801 Squadron were able to beat expert U.S. Air Force aggressor pilots flying F-5Es by a ratio of twenty-seven to ten, which "astonished" the Americans.[125] They have also defeated USAF F-15s based in Germany by a kill ratio of seven to one.[126] If that is not convincing, note that Sea Harriers have also beaten F-16s and F-18s.[127] As one Sea Harrier pilot put it, "We always did very well in defeating F-15s, F-16s, Hornets, and other agile aircraft because of the tactics we developed."[128] Once again the less expensive and less glamorous aircraft gets better results in combat.

The F-14, especially in its latter years, was also a very unreliable aircraft, but a maintenance burden as well. This plane was truly a "hangar queen." Actually, "On average, an F-14 requires nearly fifty maintenance hours for every flight hour." Jon Lake wrote in the Winter 2001/2002 issue of *International Air Power Review*,

> Experience showed that there was a strong likelihood that in any given strike package, at least one F-14 was likely to abort for technical reasons, while Hornets tended not to, while the maintenance hours required to sustain a single Tomcat sortie remained very high, making it difficult for F-14 squadrons to maintain a high sortie rate, and requiring their maintenance departments to be large and well manned. Finally, even with LANTIRN [low-altitude navigation and targeting infrared for night] and all of the other upgrades, the Tomcat was not as versatile nor as effective as the Hornet, and offered no advantages, except where payload/range was an issue or where an extra pair of eyes was relevant.[129]

Sometimes the maintenance problems in the F-14 squadrons were compounded because of low-quality personnel. Wilcox wrote that when Cdr. Joseph Aucoin, USN, took command of VF-41 in 1998, the squadron was in terrible shape. "Bad management had caused a deterioration in the quality of jet technicians and their noncommissioned managers, which had resulted in the squadron's airplanes being mostly grounded and in need of repair. The situation had gotten so bad that the squadron's previous cruise had been scornfully dubbed 'the Love Boat Cruise' by its aviators because of the lack of flying time the broken airplanes had caused them. The nickname referred to the extra time in ports they had because they couldn't fly. . . . Morale was low. Readiness was shot."[130]

The F-14 has now faded into the pages of history, and the F/A-18E/F Super Hornet is replacing it. While certainly much newer than the F-14, some say the Super (expensive) Hornet is no improvement over the existing F/A-18C/D, that in fact in many parameters it is actually *less capable* than its predecessors. Critics have roasted the new aircraft for its compromised "do it all with one platform" philosophy, and in 1999 the U.S. Marine Corps even stated that it would flat-out refuse to buy the aircraft. Even compared to the stylish but overpraised F-14, the ill-regarded and oversold Super Hornet falls short in key areas. Consider payload and range, for example. Said Bob Kress and Rear Adm. Paul Gillcrist, USN (Ret.), in 2002, "Though it's a whizzy little airshow performer with a nice, modern cockpit, it has only 36 percent of the F-14's payload/range capability. The F/A-18E Super Hornet has been improved but still has, at best, 50 percent of the F-14's capability to deliver a fixed number of bombs (in pounds) on target. This naturally means that the carrier radius of influence drops to 50 percent of what it would have been with the same number of F-14s. As a result, the area of influence (not radius) drops to 23 percent!"[131]

"The Super Hornet program is still not the performance champion among combat aircraft," echoed another critic, Bill Sweetman, in 2004.[132] "The F-15 and Rafale will carry more weapons and fly farther, and the Rafale, F-16, and Typhoon will out-accelerate and outmaneuver the F/A-18E/F at high speeds."[133] Stan Crock points out that a great many naval aviators appear to be quite unimpressed with the new airplane and consider it a step backward, not forward: "If the Joint Strike Fighter dies," frets one airman, "we're stuck with the Super Hornet."[134]

One cannot talk about modern Navy fighters very long without bringing up the movie that made them famous. *Top Gun* is just a movie, clearly, but it was made with the full cooperation of the U.S. Navy, which then exploited its popularity to boost recruiting. "Indeed the Navy liked the film so much that Navy recruiters set up recruiting booths inside some theaters that were showing the film. According to the Navy, recruitment of young men wanting to become naval aviators went up 500 percent after the film was released," said David L. Robb in his book *Operation Hollywood: How the Pentagon Shapes and Censors the Movies*.[135] Thus, it is probably fair to say that this film and the perceptions it created were quite influential in the United States. This movie was very much directed by the Pentagon, but as Robb argues, most Americans "have no idea that the government has any say whatsoever in the content of films and TV shows."[136] But it does. In order to make use of Navy aircraft, bases, personnel, and such, movie producers must surrender some of their creative license to Navy officials who act, in effect, as script censors, removing scenes and dialog that might reflect poorly on the U.S. Navy. When it comes to Hollywood productions, the Pentagon will indeed "change the facts to make the military look better than it really is."[137]

In *Top Gun* the Navy nixed early story ideas about a midair collision and told the producers to change Kelly McGillis' dishy and toothsome character (Charlotte) to a civilian so it would be safe for the Tom Cruise character (Maverick, a naval aviator) to court her. Curiously, though, several not-at-all favorable elements were left untouched and uncut by the Navy. Maverick, in his white Navy uniform, follows Charlotte into a ladies' room at the WOXOF Lounge at Miramar Naval Air Station and propositions her. Maverick, a brilliant flyer with somewhat of an Icarus complex, is depicted as emotionally unstable, unreliable, glory-seeking, sophomoric, insubordinate (his wearing of that colorful leather flight jacket off base was against regulations), undisciplined, ungentlemanly, immature, reckless, overly aggressive, and "dangerous and foolish." His trusty radar interceptor officer and sidekick, Goose, is more mature and responsible than Maverick and serves as the voice of reason, but he is killed off for dramatic purposes. Despite these massive character flaws, Maverick is only slapped lightly on the wrist for his many, many transgressions and ultimately becomes a hero for his unyielding and unpredictable flying skills. The U.S. Navy apparently had no problem with this. No matter what Maverick tried to do, he always got away with it and was even rewarded. To some cynics out there, this sounds reminiscent of the USS *Vincennes* incident, in which members of the crew received citations for shooting down an Iranian airliner. (The movie/propaganda film also put across that the U.S. Navy had a twelve-to-one kill ratio in air-to-air combat during the Korean War. Very impressive, but of course it did not mention that the Navy did very, very little in the air-to-air domain and claimed only seventeen kills.[138] The Navy actually concentrated more on close air support missions, mostly leaving the U.S. Air Force and others to handle the North Korean, Soviet, and Chinese MiGs. Along with Marines, the Navy lost more than a thousand aircraft in combat and in accidents in Korea.)[139]

Unhappily, the relationship between *Top Gun* and the Navy goes far beyond recruiting and aerodynamic showmanship, however, because, as Oscar Wilde once said, "Life imitates art far more than art imitates Life." A case in point, says Diehl, is the behavior of real Navy fighter pilots after *Top Gun* became a hit:

> Navy Secretary John Lehman (himself a party-loving Navy flier) was once photographed congratulating Tom Cruise, star of the blockbuster movie Top Gun. In the film, the Tom Cruise character flies the F-14 Tomcat at the Navy's elite fighter weapons school. Cruise's character repeatedly takes chances, such as buzzing an air traffic control tower. He enjoys being called dangerous and flies "at the edge." Unfortunately, the real Tomcat fliers would try to emulate this devil-may-care Hollywood image. The F-14 mishap rates would more than double in the years following the movie's release—in numerous not-so-great balls of fire, to borrow from the Jerry Lee Lewis song in the soundtrack.[140]

There are other things one will not see in a Pentagon-controlled movie like *Top Gun*. Take the French navy, for example. Time and again, the French military and its technologies are completely and utterly disparaged in the United States. Be that as it may, the French navy recently also scored some points against U.S. Navy fighters. In December 2002 a French magazine reported that Rafales from the much-derided new carrier *Charles de Gaulle* tangled with American F-14s and F-18s from the USS *Theodore Roosevelt*, and the combat-proven U.S. Navy planes got their money's worth out of the novice French fighter: "According to the 12F pilots, the low-speed maneuverability of the Rafale surprised their American counterparts. 'Results were positive,' modestly adds Lieutenant Commander (Philippe) Roux. . . . 'Our training focused on close combat, which emphasizes the Rafale's maneuverability, including rate of turn, turning radius, acceleration, deceleration and vertical maneuvers. . . . The Rafale proved superior in each of these areas. The American F-14s and F-18s are proven, high-performance machines, and their crews know them inside-out, but they are still previous generation planes."[141] Another French publication also noted in 2004 that "the results of engagements against other fighter aircraft used by allied countries exceeded the highest expectations. According to the pilots of 12F Squadron, the Rafale is indeed an exceptional aircraft, and the score of victories obtained against F-14 Tomcats, F-15 Eagles and F/A-18 Hornets is remarkably high."[142]

U.S. naval aviation is truly "up the creek without a paddle" these days, even against the allegedly inferior French military, but the fact is that it has had serious deficiencies for many years. In 2000, Burns declared, "We are a much less effective force than we were seven or eight years ago."[143]

> At the start of the Kosovo conflict, says Burns, who at the time was stationed at the Strike Weapons Tactics School in Virginia Beach, U.S. Navy pilots hadn't been trained in using laser-guided weapons. "That's why we had such high miss rates in the opening phases of the war. We had to dispatch someone [to tutor pilots] in laser-guided bomb delivery techniques." Burns, who retired in 1999, says that when he last served on the Eisenhower in the Mediterranean, the carrier was 'undermanned' by 450 to 500 sailors. "They didn't have enough people to keep the [approach] radar fully manned at all times." If the weather closed in, he adds, someone would have to be sent down to the bunkroom to wake up a radar operator. "The Navy says operations are safe. But they aren't safe. Planes were running out of gas and they couldn't come on board." Flight training hours have been cut back so much, says Burns, that the last time his carrier fighter squadron went on deployment, its aviators were only getting 10 to 15 hours a month.[144]

In September 2000, Lt. Cdr. Steve Rowe, USNR, lamented that "during my 12 years as a naval flight officer, I took great pride in the unique contribution of naval aviation. Navy air was the nation's enabling air arm. This unique capability is arguably no longer credible today. And will almost certainly become a paper tiger in the near future. Why? Because the leaders of naval aviation and the Navy as a whole have forgotten what the Navy is about. In the mad rush for dollars in an underfunded military, the leaders have neglected our core competencies, and grossly unbalanced support and force protection capabilities to favor strike aircraft."[145] In so doing, as noted earlier, the Navy has gutted its ASW aviation assets, but one might disagree with Rowe's statement that the U.S. military is underfunded. How is this possible, when the U.S. military budget will soon be equal to the military budgets of every other country in the world *combined*?[146] What this suggests is that even outspending the rest of the world is not enough if the money is not spent wisely. The cuts in flying hours have continued, even after President George W. Bush dramatically increased the defense budget. As Capt. Bob Scott, Supply Corps, USN, said in 2005: "The Navy flew 17 percent fewer hours in fiscal 2004 compared to fiscal 2003 and has forecast slightly fewer flying hours in fiscal 2005."[147] Without a doubt, the wars in Iraq and Afghanistan have been a drain on the budget, even though so far the United States has not lost very much equipment or very many personnel in either of these theaters, at least compared to what it lost in Vietnam, where *thousands* of U.S. planes and helicopters were destroyed and almost sixty thousand U.S. personnel died.

Given the multiple Achilles' heels already documented in U.S. naval aviation, it may not be terribly surprising that a few contemporary naval aviators now bestow great esteem on their traditional rival, the Air Force. The perceptive Cdr. Bob Norris, USN, flew F-18s in the Navy and also did a three-year exchange tour in the USAF, flying F-15s. When asked if an aspiring fighter pilot should go to the Naval Academy or the Air Force Academy in Colorado, he actually encouraged the prospect to consider the Air Force. Incredibly, for a naval officer, Norris was quite enthusiastic about the service: "The U.S.A.F. is exceptionally well organized and well run. All pilots are groomed to high standards for knowledge and professionalism.... Their aircraft are top-notch and extremely well maintained.... Their enlisted personnel are the brightest and the best trained. The U.S.A.F. is homogenous and macro. No matter where you go, you'll know what to expect, and you'll be given the training and tools you need to meet those expectations." The Navy, he said is "heterogeneous and micro.... Your squadron is your home; it may be great, average or awful. A squadron can go from one extreme to the other before you know it.... The quality of the aircraft varies directly with the availability of parts. Senior Navy enlisted are salt of the Earth; you'll be proud if you can earn their respect. Junior enlisted vary from terrific to the troubled kid the judge made join the service.... The quality of training will vary and sometimes you'll be in over your head."[148] The only truly positive aspect of flying in the Navy, according to Norris, was that "you will fly with legends and they will kick your ass until you become a lethal force."[149] So, in Norris' opinion, unlike the those of the U.S. Navy, Air Force training and aircraft are consistently excellent, and

USAF enlisted people are better trained than Navy personnel. "Bottom line, son, if you gotta ask . . . pack warm & good luck in Colorado."[150]

Maj. Gregory Stroud, Arizona Air National Guard, a former Navy pilot, "jumped ship" to fly F-16s in the Air National Guard in 1988, and he too was less than exuberant about naval aviation. Major Stroud has the great distinction of graduating from both the Navy Top Gun course and the Air Force Fighter Weapons School, and his comparison of the two courses is not flattering to the Navy. "The F-16A School (Air Force) was a much more comprehensive and difficult school which takes five months to complete and covers every tactic and mission the F-16 is capable of. . . . Top Gun was fun and easy in comparison."[151] A fighter pilot, Commander "Sharkey" Ward also points out that the legendary Royal Navy Air Warfare Instructor School was so tough as to make "Top Gun look like a holiday."[152]

Furthermore, in 2004, Captain Pedersen, a former fighter pilot, opined that Top Gun had been seriously devalued and possibly watered down. He was worried about the possible consequences: "We're making the same mistakes today that precipitated having to start Top Gun originally. We're always enamored with the latest, the greatest, the whiz bang airplanes. But we're numerically inferior again. We've fought a couple of wars where we've been unopposed in the air. Smart weapons are the thing. Everybody is banking it will be that way forever. I'm banking on China. China will bring 6,000 airplanes into play in a single day. Then what will we do?"[153]

This news about Top Gun being overrated will surprise a lot of Americans, as will the disclosure that in 1999 the finest air combat training facility in the United States was not at the Top Gun school, nor any of the other naval air stations, for that matter. No, that honor actually went to the Air National Guard's Combat Readiness Training Center in Alpena, Michigan. According to James McCrone, of *National Guard* magazine, the technology provided to the part-time airmen of the Air National Guard at this range would have made the movie characters Maverick and Iceman quite jealous.[154]

In all these cases though, U.S. naval aviation versus the Canadian navy, the Canadian air force, the Israeli air force, the Chilean air force, the French navy, the Royal Navy, the U.S. Air Force, and the U.S. Air National Guard, we can see good reason to cast doubt on Tom Clancy's judgment that "nobody else—not the Israelis, British, not even our own U.S. Air Force—makes better pilots than the U.S.N. . . . [T]he Navy trains fliers with basic flying and combat skills that are unsurpassed."[155]

11

Lack of Training, Overrated Technology, Bad Policies, and Technocratic Leadership

DESPITE ITS VASTLY SUPERIOR numbers, resources, and expensive weapons, the U.S. Navy, the world's only true heavyweight navy, continually fails to vanquish welterweight and lightweight naval powers. This would indicate that training and the selection and development of officers into professionals—not big, expensive ships and bloated budgets—are the keys to naval power. It is the personnel system—the foundation of military culture—that truly undermines the performance of the U.S. Navy, reaching out and affecting all other institutions that constitute a military organization.

For starters, an obsession with "filling spaces with faces" while abiding to politically correct personnel policies and regulations diminished training. In the early 1950s, "boot camp was supposed to take thirteen weeks, but the revolving door speeded the process to eleven, ten, or even nine weeks," says Michael Isenberg, author of *Shield of the Republic: The United States Navy in an Era of Cold War and Violent Peace*. "By comparison," he noted, "the much smaller but highly professional Canadian Navy devoted twenty weeks to 'preentry' training."[1]

A more recent example concerns submarine training. Though the United States maintains the largest submarine fleet in the world (now that the Russian fleet is mostly tied up at dockside), American submariners do not currently receive escape training. The Canadian submarine force has only four boats, yet it has managed to acquire the most advanced submarine escape training facility in the world. In November 2003, the U.S. Navy was considering sending its submariners to Canada for escape training.[2] (According to Karam, the U.S. Navy closed its free-ascent training facility, but even if it had been open during his tour of duty in the late 1980s, he, an enlisted "nuke," would not have received such training, because the U.S. Navy apparently felt that only the non-nuke sailors, and the officers, needed it.) One might think it is rather strange that the richest kid on the block was actually thinking about visiting his poor cousin to go swimming. The British too have an excellent submarine escape training facility, and after touring that facility a few years ago Truscott remarked, "It is hard not to be impressed with the quality of training offered to British submariners."[3] One can only imagine his reaction if he knew that American submariners at that time did not receive this kind of training at all. I am happy to report, however, that the U.S. Navy will indeed get back into the submarine escape training business, but not until 2008, when its Mark 10 Submarine Escape Immersion Equipment (SEIE) will come on line. When it does, "'This SEIE facility will put the U.S. Navy back on par with the level of escape trainers now found in England and Australia,' said Lt. Joel McMillan, assistant Resident Officer in Charge of Construction."[4] It is hard to believe that what some call "the best navy in the world" cannot currently provide realistic training on par with that of "lesser" naval powers like Australia.

In response to these criticisms, some U.S. Navy boosters will say, defensively, that the number of days spent under way ("steaming days") per year is strongly correlated with the overall combat readiness of a navy and that by that standard the U.S. Navy does very well—and many agree. Yet, if the number of steaming days in itself is a valid way in itself to evaluate readiness, then probably the best-trained navy in NATO in the late 1990s was the pocket-sized Belgian navy. In 1996, a U.S. Navy officer visited a Belgian ship during exercises in the Baltic Sea and reported that "Belgian sailors seem never to stop sailing," averaging 280 steaming days a year.[5] The following year, the U.S. Navy average for deployed ships in the Atlantic was only about two hundred steaming days, and in Operation Desert Shield/Storm elderly Canadian ships detached to the Multinational Interception Force in the Persian Gulf maintained higher "on station availability" than did the U.S. Navy ships and were praised by U.S. Navy senior officers for being "an example for others to follow."[6] In 1990, Dunnigan and Nofi rated the Royal Navy's overall seamanship as superior to that of the U.S. Navy[7]—but then, that really should be no surprise. German notes that in the late 1940s, in their first commands, graduates of the Britannia Royal Naval College were generally "better seamen and navigators, better shiphandlers, and more flexible tacticians" than were Annapolis graduates.[8]

The U.S. Navy opines that its officers and crews are the most professional in the world, yet media reports have indicated that a startling number of American ship commanders have been fired or suspended in recent years, including the captain of the carrier *John F. Kennedy* after his ship collided with a small dhow in the Persian Gulf in 2004. Accidents happen in every navy, but in his discussion of why so many U.S. Navy commanders are getting fired, Raymond Perry has said, "I believe that the spate of CO firings is an indicator of the decline of professional warfighting skills of naval officers."[9] Perry, who served for twenty-nine years as an officer, said that when he was at the U.S. Naval Academy in the 1950s, one of his professors observed that "operational competence was no longer a true priority in the U.S. military."[10] Officers become so engrossed on how to play and master the rules of the promotion and selection system that the attainment of enlightenment of the rigors of the military profession is subordinated, even frowned upon. No one wants to know more than their superiors.

Rear Admiral Jeffry Brock, RCN (Ret.), verified this assertion in reminiscing about an exercise he observed in the 1950s. During the initial phase of the exercise, he reported that the U.S. Navy admiral (and staff) commanding the U.S. Second Fleet, while en route to the United Kingdom, "lost almost two-thirds of their own forces. Furthermore, the exercise referees concluded that most of this damage to the United States Atlantic Fleet had been brought about by disastrous mismanagement and the misdirection of their own attack forces."[11] Brock also let on that even before losing most of the U.S. ships in simulated combat the American admiral had asked him to arrange discreetly for a Canadian ship to "pass across one or two charts of the northern-west approaches to the United Kingdom and the entrance to the Clyde."[12] (Note that this happened *after* the fleet was already under way.) Brock obliged, and soon thereafter he observed something peculiar—helicopters from the *Saratoga* were frantically delivering hand messages to the rest of the U.S. Second Fleet. Then someone quietly informed him of what he had already suspected, that "the whole United States Atlantic Fleet had sailed without proper charts of their destination"[13] and that the Canadians had saved the day. Brock concluded, laconically, "I was appalled." He also said the

American version of the postexercise review was, to put it mildly, "a white-wash."[14]

That was almost fifty years ago, and perhaps not much has changed since (recall the Navy's attempt to deny that it was defeated during Millennium Challenge 2002). Both the professor and Perry argued that then and now, political maneuvering and impressing the brass take priority over war-fighting skills in the peacetime U.S. Navy.[15] Many conscientious officers quietly agree with them. The reason why stultifying careerism runs rampant among U.S. Navy officers is directly related to several components of the Navy's personnel system (in reality, the U.S. military's personnel system). Four pillars make up that system, or as Don Vandergriff calls it, "the pillars in which the military culture rests upon":

The "up or out" promotion system
A bloated officer corps
Individualistic, focused, and systematic evaluation systems
Centralized promotion and selection boards.

Today, the Navy, as well as the other services, manages personnel under the Defense Officer Military Personnel Act of 1980, or DOPMA. This law centers around a promotion approach called "up or out." This method, enacted in 1916, is used by the Navy, and in all branches of the U.S. military, to manage its leaders in the information age.[16] This system, unlike those used in other English-speaking navies, requires U.S. Navy officers (and enlisted personnel) to be promoted "on schedule" or face early retirement or separation. This in turn creates insecurity, competition, and a desire for impossible perfection, which in turn encourage dishonesty and a zero-defects mentality, which applies to everyone except the crew of the USS *Vincennes* (more on this in a minute). According to Vandergriff's upcoming book *Raising the Bar: Creating and Nurturing Adaptability with the Changing Face of War*, the "up or out" system is "first and foremost" of the pillars, and, combined with the other three, it prevents Navy officers from becoming "adaptive leaders," capable of innovation and the ability to fight new, nonlinear threats. He argues that if the Pentagon can "knock down" the up-or-out policy, the other three pillars will face the same fate and thus allow the Navy to change its culture and allow it to accommodate and adapt to new threats efficiently. It would allow officers to focus more on their war-fighting skills than on their careers, and this in turn would require less ticket punching, which has assumed almost absurd levels.

This American ticket punching bemused Brock during the Korean War:

> I was intrigued by the frequency with which the command and fleet organization structure would be changed. I was eventually forced to the conclusion that much of this was due to an American desire to give as many senior officers as possible what they called 'battle command experience.' But it happened with such regularity and for such periods that it accomplished little for the war effort except to confuse this peregrinating band of heroes as much as it did the rest of us. Furthermore, there was no enemy at sea to provide the 'battle experience.' I believe, however, that these movements of U.S.N. ships and Flag Officers in and out of the war zone also had something to do with the kind of paper records needed for promotion—to say nothing of the acquisition of more medal ribbons, for which there was keen competition.[17]

Actually, as time went by, the competition for medals eased considerably, thanks to increasingly liberal standards. For example, during the Vietnam War, "medals eventually produced results contrary to their intended purpose. Because awards were made almost on a production-line basis, the value of decoration declined. In a war where computer programmers received Bronze Stars, other decorations—bravely earned—were diminished by comparison. So many sorties for an Air Medal, certain results for a DFC [Distinguished Flying Cross]; a Silver Star for a MiG kill. The latter rankled some hardcore fighter pilots. After all, how many fliers in World War II received that decoration for shooting down one airplane?"[18]

Interestingly, though, the desire for battle experience and medals is only strong when they are perceived as glamorous and/or career enhancing. When Admiral Zumwalt was Commander, Naval Forces Vietnam in the late 1960s, he oversaw the brown-water campaign on the rivers. Zumwalt was utterly dismayed at the poor quality officers that seemed to be prolific in his command. He wrote that "the level of competence hovered around zero. Vietnam was a dumping ground for weak naval officers at the commander and captain levels."[19] At the headquarters in Saigon, "many officers were more concerned about their tennis dates and dinner plans than putting out a maximum effort to fight."[20] As for the reason why this was so, one need consider only thing: "Obtaining good officers was difficult since many good naval officers perceived that Vietnam service would be of no help to their careers, and more likely a step backward. I had always thought that when your country was at war, you sent your best men to fight it. I knew there were personnel detailers in Washington telling good naval officers not to go to Vietnam, and offering them a year at the Naval War College instead. In effect, these military officers were dodging the war just like the young men who ran off to Canada."[21] To put it another way, a tour in Vietnam might be a ticket to "out" rather than "up" in the minds of these careerists. This "up or out" system also ensures that some of the Navy's most experienced and mature officers will be lost into the civilian world, because, after all, only a very few officers in any navy will ever make flag rank. Even former Defense Secretary Rumsfeld trenchantly called the U.S. Navy's officer promotion policy "lousy" in 2003, yet it remains in place at the time of writing.[22] This is a systemic problem.

In addition, one should also recall the attack on the USS *Stark* and the shoddy damage control procedures used by her crew (men were fighting fires while wearing improper clothing, and there was not enough equipment, although this was not the case, thankfully, when the USS *Samuel B. Roberts* struck a mine in 1988); the accidental and inexcusable attack on an Iranian airliner by the USS *Vincennes;* and the more recent collision between the nuclear submarine USS *Greeneville* and a Japanese vessel. In the latter case, when the Japanese government found out that untrained civilian guests had actually been at the controls of the *Greeneville* before the collision, they were most undiplomatic. "It is outrageous. The U.S. Navy is slack," said the Japanese Defence Agency Chief, Toshitsugu Saito.[23] Also note that the Japanese have gone through this before: "In 1981, the nuclear submarine USS *George Washington*, en route to a liberty port, hastily surfaced in the East China Sea. It rammed and sank a Japanese freighter. Unbelievably, the sub did not report this collision until the next day."[24]

Paul Beaver, military editor at *Jane's Defence Weekly*, told National Public Radio's Lisa Simeone in 2001 that the U.S. Navy is quite probably the only navy in the world that has a "civilian ride-along program." Although civilians can visit British and Canadian warships, for example, they may do so only when the ships are at dockside; they must leave the ships before they get under way. He added that Britain's Royal Navy would never even consider such a ride-along program, because of the inherent risks involved.[25]

Regarding the *Vincennes* incident, former *Chicago Tribune* military correspondent Lt. Col. David Evans, USMC (Ret.), said it was "an operationally inept tragedy that caused the loss of 290 civilians, when the skipper had electronic [transponder] evidence that the 'target' was not an Iranian F-14 but a commercial airliner, not to mention that the captain was in Iranian territorial waters, where he had no business being since he was not under attack. Many U.S. Navy officers feared this sort of thing could happen, calling their apprehension a case of 'Aegis arrogance.'"[26] Even though nearly three hundred innocent civilians were killed, the captain of the *Vincennes,* who also ignored a direct order to hold his position, was soon decorated with circumstantially dithyrambic praise in the form of the Legion of Merit.[27]

Shuger states that the *Vincennes* tragedy was bound to happen sooner or later: "During my own experience in the Navy from 1979 to 1983, I repeatedly found defects in the Navy's planning and preparation, defects that were individually exasperating and, collectively, indicate that the mental confusion on the *Vincennes* was especially severe but not uncommon."[28] He also says that this incident "offers an illustration of how the Navy's readiness problems stem from human, rather than mechanical, deficiencies,"[29] and as an example he offers the following: "In reacting to what it took to be an Iranian F-14, the *Vincennes* crew displayed inadequate knowledge of a U.S.-made threat aircraft. Captain [William C.] Rogers and the crew thought they were aiming at an American-made plane, yet from what we know about what happened aboard, the crew displayed remarkably little understanding of how the F-14 works."[30] To wit, "The F-14 is a fighter/interceptor designed to shoot down fighters and cruise missiles. . . . In other words, it is not a bomber and the odds of one of its Sparrow missiles successfully looking down into the electronic clutter produced by any body of water and locking onto a surface target are as low as the odds of a torpedo hitting a low flying-plane. In addition, nobody on the *Vincennes* seems to have noticed that or wondered why the ship's radar-detecting equipment didn't spot any sweeps of the unique F-14 radar coming from the blip."[31] Nevertheless, the airliner was targeted and destroyed.

After the USS *Vincennes* outrage, many agreed that the impenitent U.S. Navy was fundamentally flawed in a number of cogent areas, and many wondered about its so-called wonder technology, as well as the training deficiencies outlined above. Actually, many believe that the U.S. military's claims of a vast "technological edge" over other countries are lacking in substance. Capt. Larry Seaquist, USN (Ret.), said in the 1993 book *War and*

Anti-War that the United States "has no technological monopoly in virtually anything . . . I've never found anyone to respond to my challenge to name three technologies which are under the exclusive control of the U.S. military. There's nothing left."[32] In their controversial 1991 book *The Coming War with Japan,* authors Friedman and Lebard argued that "the Japanese are in the forefront of high technology maritime construction," that Japanese destroyers are the equals of U.S. Navy ships (of course, Japan had access to U.S. technology to accomplish this, though it is rather unlikely that the United States would sell Japan sensitive military technology unless it knew the Japanese already had the wherewithal to develop it themselves), and that "in certain technologies, such as electronics miniaturization—useful in advanced avionics, and fire-control systems, Japan is ahead of the United States."[33] They also described Japan's indigenously designed Type 90 main battle tank as "the finest main battle tank in the world,"[34] asserting that with a well trained crew it would be a very dangerous opponent for the American M-1 Abrams tank.

For their part, the Russians also have a rocket torpedo, known as the Shkval, that "travels at a speed of 200 knots, or five to six times the speed of a normal torpedo, and is especially suited for attacking large ships such as aircraft carriers." As Charles Smith points out, "The Shkval was designed to give Soviet subs with less capable sonar the ability to kill U.S. submarines before U.S. wire-guided anti-sub torpedoes could reach their target."[35] If these claims are even partially true, these new Russian weapons could spell big trouble for the U.S. Navy, and plenty of others as well.

The U.S. Navy appears to be lagging behind the Russians in rocket torpedo development, but that is not the only area in which the U.S. Navy has stumbled. Indeed, as I mentioned earlier, even its communications equipment has proven less capable than that used by other navies, as early as the 1990s. Morin and Gimblett have suggested that even though Canada sent older ships to patrol the Persian Gulf during Operation Desert Storm in 1991, those ships had far better communications equipment than did the U.S. Navy, and unlike the American ships in theater, the Canadians were able to communicate easily with most of the allied coalition participants.[36] Dr. Richard Gimblett, a retired Canadian navy officer, noted in 2004 that "in private conversations, U.S.N admirals will candidly admit that the Canadian Navy manages the frigate navies of other nations better then they could hope to."[37] This undoubtedly has to do with the fact that the Canadians have superior communications systems, which the Americans frequently need. When his ship patrolled the Persian Gulf, for instance,

Captain Abrashoff noted that "sometimes the communications problems were serious, and one equipment snafu nearly scuppered the whole Gulf fleet in the Iraqi crisis of 1997."[38] Thousands of operational messages were delayed or just disappeared over several weeks, thanks to a cumbersome new satellite system, and the fact that the U.S. Navy had not even trained anyone how to use it!

A British naval officer who recently completed a two-year exchange assignment with the U.S. Navy told me in 2005 that even though he has great respect for that service, "the R.N. is an incomparably better navy than the U.S.N."[39] While Burns complained that American ships are sometimes undermanned, which is true, some U.S. Navy units have far too many personnel. Lieutenant Commander Aidan Talbott, RN, described the U.S. Navy as "cumbersome, vastly overmanned, stolidly managed, with massive institutional inertia and hobbled by internal and external politics."[40] He went on to say that he endured "monstrous levels of inefficiency in many respects" during his two year tour and that in some ways even the most modern American surface ships were technologically "old-fashioned and manpower intensive" compared to British ships.[41] "The engine control room of my last RN ship, a Type 23 frigate, was much more advanced than the *Arleigh Burke*s (DDG 51s) I went onboard. T23s and DDG51s are of (again, I think) similar design and procurement vintage, yet the *Arleigh Burke* control room was very much valve wheels and dials compared to a T23 with remote operated control and digital displays. Don't get me wrong—T23s are hardly the latest technology by any means, but it was a significant change from an *Arleigh Burke*."[42]

Not only is the U.S. Navy's technological lead over others now largely illusory, that very technology, which is often used in an attempt to compensate for poor training, can certainly be a weakness and not a strength, especially against an intelligent and better trained enemy. For example, in the late 1980s and early 1990s, the U.S. Navy, along with many other high-technology military organizations, faced a potentially debilitating supply problem related to memory chips. At that stage, the U.S. Navy had become almost totally dependent on Japan for semiconductors; as TV journalist John Chancellor put it in 1991, "These tiny chips are needed for everything from supercomputers to jet aircraft, and makers of the most sophisticated electronic equipment, including military contractors, must depend on Japan for supplies. You can't run today's world without chips, but more and more, the United States, including its military, is dependent on Japan."[43]

One Japanese politician even said that if Japan decided to sell its memory chips to the Soviet Union rather than the United States, the balance of power would have been jeopardized.[44] Thankfully for the U.S. military, the Japanese gradually lost their stranglehold on the semiconductor market, and the American industry rebounded. Notwithstanding, for a time the U.S. Navy was almost completely dependent on its former enemy for essential computer hardware, and adversaries could theoretically have exploited this supply weakness. Obviously, the United States and all its high-technology allies would have needed to maintain very good relationships with the officially pacifistic Japanese if they had to fight a protracted war during that era. A small, highly mobile, low-technology enemy who had the savvy to exploit a falling-out between the United States and Japan could have been very dangerous indeed. An unlikely scenario, yes, but still possible and plausible.

Even with a steady supply of semiconductors, the U.S. Navy's technology has not always been universally admired. In the early 1990s, for instance, it was one of more than thirty navies that visited South Africa. The chief of the South African navy kept careful notes on all the visiting warships, giving the highest points for technology and personnel to the navy that impressed him the most. The winner was not the U.S. Navy, or the Royal Navy, or any of the other usual suspects. Instead, the South African admiral gave the top honors to the modest and understated navy of America's northern neighbor, Canada, for reasons that will be detailed below.[45] This honor confirmed what many knowledgeable and worldly naval experts already knew about the Canadian navy, whose ships regularly outmatch U.S. Navy warships.

Although most Americans do not associate Canada with high-technology military or naval systems, U.S. Navy officers fawned over the new Canadian Patrol Frigates in the late 1990s, frequently stating that they were in most respects better than U.S. frigates and destroyers. In a July 1997 report by the U.S.-based Center for Security Strategies and Operations (CSSO), the Canadian *Halifax*-class frigates were compared to similar vessels from five allied nations. The CSSO argued that the Canadian ship had better self-defense systems than U.S. ships, by virtue of its unique "completely automated combat system."[46] "Of all the frigates analyzed the *Halifax* class emphasizes survivability to the greatest extent,"[47] the report declared. "The *Halifax* is the only frigate analyzed that has an advanced, state-of-the-art, fully distributed combat system with a distributed command and control system linked by redundant data buses."[48]

The Canadian ship was also rated the highest in ASW capability.[49] The American periodical *Forecast International* chimed in, stating that "the *Halifax* class frigates have matured into fine warships. The lead ship of the class has been the subject of unstinting praise from the U.S. Navy, following visits to American naval bases."[50] In 2004, Wajsman recorded that the Canadian ships have a communication system that is "the envy of many NATO countries. It allows calls to be made from a compact console to anywhere on the ship or to anywhere in the world at the touch of a button. And it operates with multiple inter-face capability."[51] He also said that the Canadian Zodiac fixed-hull light boats are the finest in the world and that the CANTASS system remains "highly regarded" in the international arena.[52] Gimblett also mentioned that during naval operations against Iraq (2001–2003), "the signal processing and tracking capabilities of the Canadian frigate combat systems were proven to be superior in some respects even to the nominally next-generation features of the *Aegis* system."[53] Indeed, it turns out that the only major shortcoming of the *Halifax* class is that it does not have three-dimensional radar (i.e., capable of finding height). These capstones were a major coup for a navy with only nine thousand sailors and a 1998–1999 budget of less than $1.5 billion (U.S.).

If one compares the *Halifax*-class frigate to the U.S. Navy's premiere destroyer, the *Arleigh Burke* class, the *Halifax* arguably has some distinct advantages in ASW. For example, all the *Halifax*-class ships have a world-class towed array sonar system, whereas only the first flight of the *Arleigh Burke*–class ships have a towed array. All the *Halifax*-class ships have an embarked ASW helicopter (the CH-124 Sea King), which, although quite old and in dire need of a replacement, is large, fully autonomous, and therefore able to search for and attack submarines independently. Only the updated version of the American ship (the Flight IIA version) has embarked ASW helicopters, but the Americans prefer not to use their ASW helicopters as autonomous assets but rather as tethered extensions of the mother ships. For ASW, the U.S. Navy LAMPS III SH-60 Seahawk helicopters relay acoustic data back to the ship for processing and receive operational directions from the ship through a datalink, which in wartime could be vulnerable to failure, jamming, or spoofing. As a result, the U.S. Navy LAMPS III helicopter crews are quite limited in their ability to take the initiative and are not true "force multipliers" like the autonomous helicopters of the British and Canadian navies. Interestingly, although the Canadian Forces make great efforts to remain interoperable with their American brethren, the U.S. Navy Seahawk failed to impress

Canadian officers during the recently completed competition to replace the Sea King. With some exceptions, Canadian senior officers much preferred the European-designed EH-101 Merlin over the American models, but they were compelled to accept the Sikorsky H-92 (a larger and improved version of the Seahawk) because the more powerful and capable Merlin was too expensive.

Talbott also noted that U.S. Navy officers were much more familiar with the technical aspects of their jobs than the principles of good leadership, and he felt that their training was "excessively narrowly specialized" and "extremely stove-piped."[54] He suggested that U.S. Navy surface ship sailors are usually not trained to multitask as well as their peers in the Royal Navy: "A British sailor would be trained to multitask more, running X, W and Z systems whereas my experience was that a U.S. sailor would only be trained to manage Y system. Note that I say it is only what they are trained to do—they may well be as capable given the equivalent training."[55] Captain Abrashoff confirmed this in his 2002 book, *It's Your Ship,* in which he noted, "When I took command of *Benfold,* I discovered that the usual policy was to have only one crew member able to perform each job: one job, one person. As a result we were one-deep in just about every critical position. In effect, I was held hostage by the key people on the ship. If they left for any reason, I would have to scramble to get the job done, and probably not done well."[56] This system may be common in the U.S. Navy, but it certainly is not in many other navies, and when Abrashoff ordered his crew to cross-train and learn how to multitask, as sailors in other navies do, they resented him.[57]

Even the U.S. Army, which is also inefficient and oversold, is better at multitasking than the Navy. In 2003, Nate Orme reported that the soldiers manning the U.S. Army's experimental heavy-lift catamaran were not like the Navy's sailors. He said that "Army engineers have to be jacks-of-all-trades. Unlike the Navy, which has a specialist for nearly every task aboard a ship, Army sailors must multitask, since the crew size, about half that of a comparable Navy vessel is small and operational doctrine is still being written."[58] Canadian sailors have also proven to be much more versatile than their peers in the U.S. Navy. For example, while the U.S. Navy has specialized personnel, such as SEALS, for conducting opposed boardings, "Canada's medium navy must resort to multi-tasking its sailors. The hidden benefit is that they are always there in the ship. Non-compliant boardings resisted by the target vessel normally would be left to the specialist teams, but in what would prove to be the last Operation Augmentation deployment, HMCS *Winnipeg* was the only ship in the USS *Constellation* carrier battle group with such a capable team. She undertook a total of seven of them, with a couple being especially dangerous," noted Gimblett.[59]

It is also notable that it is not just U.S. Navy sailors who are overspecialized; some of their ships are too. This is especially true with replenishment vessels. During Operation Enduring Freedom, the Canadian replenishment ships in the Persian Gulf were "especially popular with the U.S.N because of the Canadian style of 'one-stop shopping' with all kinds of fuel and stores available from the same vessel instead of the American style of a specific ship for each, requiring a long succession of replenishments from the full fleet train."[60]

Many others have also cited overmanning and overspecialization in the U.S. Navy, and the Blue Angels are sometimes criticized for this fault. The U.S. Navy boasts that the Blue Angels flight and maintenance teams are the world's best, but when one examines their very large, seemingly bloated, maintenance team, one really has to wonder. The Blue Angels perform with only six modern single-seat F-18 jets, whereas the Canadian Snowbirds fly nine two-seat Tutors, which are very much older. In an average year, each team usually does the same number of performances throughout Canada and the continental United States. (According to the Snowbirds' official Web site, "On average they will fly approximately 70 air shows at fifty different locations across North America," whereas in 2004 the Blue Angels were scheduled to fly seventy shows at just thirty-four locations.)[61] The Canadian team flies more airplanes and has only two spare planes (the Blue Angels have three), but it still manages with a much smaller maintenance team at each show. When the Blue Angels pilots do an air show, they bring along an additional thirty-five to forty personnel in their very own C-130 Hercules, which, for no practical reason I can think of, is custom-painted in the same blue and gold scheme as the F-18s, but the Snowbirds need to bring along only about ten or eleven people, and obviously they do not have nor need their very own "airliner." (Of course, both teams have more personnel at their home bases.)

Clearly, the two-engine F-18 is a more complex aircraft than the forty-year-old, single-engine Tutor, and that might partially explain the manpower difference (although this explanation failed to convince fully some of the Canadian aviation technicians I interviewed). It is also true that U.S. Navy technicians are very specialized, and as a result they need lots of them to do the same jobs that one Canadian technician can do alone. Said Master Corporal Frank Gough, Canadian Forces (U.S. equivalent E-4) in

1993, "We have only five major trades which work on the aircraft. We have people who are more diverse and they can work on many systems at the same time."[62]

By the mid-1990s, the Canadians had only three major aircraft "maintainer" trades, based on land and on destroyers, whereas the U.S. Navy had seven comparable ratings. (see table).

Corporal Wes Cochrane (U.S. equivalent E-3), an Aero-Engine Technician (now reclassified as an Aviation Technician) in the Canadian Forces, told *Air Force* magazine that U.S. Navy aircraft technicians are awestruck when they meet Canadian technicians and compare skill sets and training: "They're surprised when they hear me list off the systems that I'm qualified to work on: the engines, drive-train, fuel systems, flight controls, hydraulic systems, and so on. They're quite amazed."[63] In 2005, Canadian air force training curriculum developers surveyed and compared Canadian courses with all the other military aircraft technician training programs in the United States and found that the Canadian Aviation Systems Technician course provides "the most comprehensive basic aircraft technician training in North America today."[64] Withal, some Canadian Forces aviation technicians have said that a trained Canadian private (U.S. equivalent: E-2) is in some ways comparable to an American E-6 in terms of their knowledge of aircraft systems.

Thus the Canadians are practically omnicompetent, while the U.S. Navy training system produces disproportionately large numbers of specialists with relatively shallow and compartmentalized training, as does the rest of the U.S. military. If the Blue Angels ground crew were really trained to multitask, as Canadian techs are, there would probably not need to be thirty-five or forty technicians at each performance. Nor would they need fifteen crew chiefs, when the Snowbirds have only one. This does not sound like an efficient or cost-effective arrangement, to say the least, and it may have something to do with U.S. government's policy of using the services as "job training" or "make work" platforms for the economically disadvantaged. In other words, the Navy and the rest of the U.S. military often employ more people than they really need so as to provide more employment opportunities to the youth of America. Lt. Jason Hudson, USN, said in 2003 that the U.S. naval service should not be a "remedial social program" for troubled or unemployed youth, but it frequently is, nevertheless.[65] As for the Snowbirds and Blue Angels aircrew themselves, both are world class and very highly trained, although diminished standards were clearly evident in the Blue Angels in the 1990s. In 1996, the commanding officer of the Blue Angels, former F-14 pilot Cdr. Donnie Cochran, USN, resigned because he did not feel that his flying skills were up to the task. Other Blue Angels pilots described Cochran as "a solid but not outstanding pilot who was not of the caliber needed to excel in the extraordinary maneuvers for which the team is famous."[66] Remember, Cochran was not just a Blue Angels pilot: he was the CO, and as such, he *should* have been the most experienced and skilled pilot in the squadron. He was not, and it does not reflect well on the Navy or its training and selection systems that he was nevertheless selected to lead the Blue Angels.

Overmanning affects not just the enlisted personnel but younger officers as well. In 2003, Lt. Kevin M. O'Neal, USNR, stated that there were supposed to be only eighteen officers on his frigate but there were in fact thirty-eight. In his words: "This situation has serious ramifications. Division officers are showing up to no jobs. On day one, they lose faith in the system. They lose the incentive to work hard because they know the system is going to ask very little of them. 'What is my job going to be?' 'You're going to be the assistant safety officer.' 'What does that entail?' 'I don't know; you're the first one. Oh, we don't have a place for you to sleep either.' Welcome aboard."[67] This means that many officers will be denied the training and experience they need to become leaders because there are simply too many officers and not enough real jobs for them. Ironically, this is not an accident, either. This overmanning is actually the by-product of a policy intended to improve officer retention. In other words, instead of trying to improve retention, the Navy just commissions more officers than it would otherwise need! O'Neal summed up, "Leaders of the surface warfare community must address this serious issue. We need to reduce first-tour division officer manning on surface ships. More people does not equal better product. I am failing my junior officers. This is not acceptable."[68]

Overmanning was also a concern for Shuger, who served in an E-2 squadron (VAW-116) aboard the carrier USS *Constellation*. Even though intelligence officers are not treated with great respect in the Navy, every carrier seems to have far too many of them.

> Each squadron had at least one air intelligence (AI) officer and the medium bomber squadrons had two. And the airwing commander has his own intel guy. On top of that aviation complement, the ship had its own intelligence division of several dozen enlisted specialists headed up by a commander and staffed by four or five other officers. And the carrier

task group commander (a two- or three-star admiral) had his own intel staff—usually several commanders and/or lieutenant commanders. On paper, our job was to manage information about any potential task group enemy. But in reality there was rarely enough enemy to go around.

Shuger attributed this "intelligence bloat" to "pointless empire-building,"[69] and I think many would agree. It clearly degrades individual training and experience, as O'Neal said, and these are key elements in combat.

One does not like saying this, but poorly trained and inexperienced personnel have made regular appearances in many U.S. Navy units for many, many years. For example, at one point in the 1950s, "over 40 percent of all naval officers had less than four years of commissioned service."[70] More recently, Williscroft said the crew of the USS *Independence* was "poorly trained" in 1998, and this, combined with broken-down equipment, impaired her readiness.[71] Capt. Ronald H. Henderson Jr., USN, who took over command of the *Kennedy* after she failed her INSURV readiness inspection in 2001, had harsh words for some former members of the ship's company, especially the chief petty officers and officers. "It was clear to me that there were a few chiefs in *Kennedy* who were, in fact, incompetent. But there were a lot of chiefs who weren't getting any support from the chain of command,"[72] he noted. His predecessor, Capt. Maurice Joyce, USN, had been relieved of his command, as was the ship's chief engineer. "What makes me really upset is when we make the same stupid mistake over and over again," admonished Henderson.[73]

Stupid mistakes are easy to make when a ship has an inexperienced and undertrained crew. Said Henderson of the *Kennedy* crew in late 2003, "Forty-five percent of my crew has never been to sea, ever, in any ship, on any ocean."[74] In 1999, Dorsey reported that 50 percent of the officers on the destroyer USS *Arthur W. Radford* (a ship that suffered a serious collision) were just ensigns, the lowest-ranking and least experienced officers in the Navy.[75] A 2002 study by the RAND Corporation confirmed that the experience levels of U.S. Navy personnel do not compare favorably with French navy and Royal Navy and Royal Air Force (RAF) personnel. The study compared the skills and experience of U.S. Navy F-18 pilots with RAF Tornado pilots and French navy Super Etendard aviators and found that "the British and French pilots have greater experience levels and more continuity in their units than the U.S. pilots."[76] In the same vein, a visit to a U.S. Navy F-18 squadron by defense analyst Franklin Spinney in 1994 revealed "deteriorating readiness" and alarming training deficiencies. "According to the squadron commander, pilot training (particularly for those junior officers embarking on their first cruise) was barely adequate prior to the deployment.... Furthermore, over the last 12 months of the cycle (the last six months of workup plus the six months on cruise), junior officers averaged 100 instead of the normal 120 carrier 'traps.'"[77] Another report issued in January 2000 confirmed that U.S. Navy pilot skill levels in general were declining at the Navy's air combat training facility at Fallon, Nevada. As the author noted, "Because incoming pilots are less proficient, Fallon basically uses its first week of flight training to bring pilots up to where they should be."[78]

The RAND study also compared American P-3 Orion crews and DDG 51 destroyer crews with their French and British ASW counterparts and concluded, once again, that the American ASW crews were, on average, the least experienced and the least cohesive. The French and British units were more cohesive and provided greater continuity because "while the typical career pattern for U.S. Navy officers takes them away from the operational ship world to various headquarters and staff assignments, French and British naval officers may stay in the operational community throughout their careers."[79] (The French are actually much better than most Americans think. In 1999, the captain of the USS *Halyburton*, for example, appraised French [and German] sailors as being "the consummate professionals," and applauded their "crisp radio calls and sharp ship maneuvers.")[80]

Spector noted that even Soviet officers "spent far more time in each assignment than their American counterparts and often remained in the same ship for four or more years."[81] Soviet officers also stayed in their navy much longer, on average, than their U.S. Navy peers, roughly 90 percent staying in the service for twenty years or more.[82] And back in the days when Canada still had an aircraft carrier, naval air crews manning the Tracker ASW aircraft stayed together for four years, much longer than their counterparts in the U.S. Navy. Interestingly, in this particular case, we see an exception to the standard "stove-piped" training that so often typifies U.S. Navy surface and air personnel. To be more specific, some U.S. Navy aircraft maintainers are also actually trained to serve as carrier aircrew, whereas the Canadians used fully dedicated aircrews only. "The U.S.N.," noted Soward, "did not employ such a defined crew structure, often using squadron electronic maintenance technicians to fill in as crew members when required. Obviously the Canadian system could not be easily duplicated in the larger and more complex aircrew personnel structures of the U.S.N."[83] This apparently unique long-term crew cycle allowed the Canadians to form highly cohesive units, and as a result those crews were said to be "profoundly superior" to U.S. Navy carrier aircrews in ASW.[84]

One of the main reasons for this lack of continuity is that career U.S. Navy officers are required, by law, to complete "joint duty" assignments (that is, in the other branches of the armed forces), which as Perry said in 2005, require "specific education . . . and years spent away from an officer's chosen specialty. My own experience has confirmed that this significantly reduces an officer's available time for professional development in his critical specialty."[85] Perry also said this was a key factor in the recent and nearly catastrophic accident involving the nuclear submarine USS *San Francisco*. The CO of the badly damaged ship, which collided with a seamount, he suggested, had not had "enough time on the pond" because of the joint duty obligation.[86] Here we have yet another systemic problem that interferes with training.

U.S. Navy enlisted men are also rather frequently shuffled from unit to unit, and new sailors have shorter enlistment contracts than do their counterparts in the French and British navies.[87] While aircraft maintainers in the U.S. Navy remain in their units for only two or three years, their counterparts in the Royal Air Force can remain with a squadron "indefinitely."[88] In one RAF squadron, an officer reported that some of the ground crew had been with the unit for seven years; one will not see this kind of longevity in the U.S. Navy.[89] German units stay together even longer, in some cases twenty or thirty years.[90]

The constant shuffling or rotation of personnel in the U.S. Navy was a great concern to Gabriel, who says that it prevents people from becoming experts. "Most American officers," he charges, "are amateurs. . . . Amateurism is, of course, directly associated with rotational turbulence. Officers who move frequently are just about reaching a level of expertise where they can stop learning their job and carry out their tasks effectively when it is time to move to another assignment."[91] Capt. Neil Byrne, USN (Ret.), said much the same thing in 2001, and Captain Seaquist too points out that during the course of any given year the average U.S. Navy warship will replace 50 percent of her crew with newcomers.[92] Things were even worse in the 1950s: "In the last six months of 1957 *Forrestal* had to replace 60 percent of her crew."[93] Admiral Rickover once remarked, "We should at once back off this infernal rotation of military people."[94]

The RAND study also observed that unlike the British and the French forces, U.S. Navy aviation units do not maintain consistent readiness to go into battle throughout the fiscal year. As Scott Shuger has said, "Amazingly, it's not uncommon for navy squadrons to cut back their flight hours drastically or even to be grounded due to the scarcity of aviation fuel near the end of the fiscal quarter. This even happens to squadrons already at sea. Several times during my carrier service we had to drop anchor and wait for more fuel money."[95] In 2000, an anonymous U.S. Navy officer informed Dougherty that during a recent exercise in Asia "five U.S. warships—including the flagship for the U.S. Seventh Fleet—[were] restricted from getting underway due to steaming-dollar shortfalls."[96] This inconsistent readiness is due to the U.S. Navy's rigid deployment cycle system and its "training philosophy." The authors concluded that this "readiness bathtub" has "caused concern at the Chief of Naval Operations level."[97] The French and British do not have this problem, because they do not use "fixed deployment and training cycles," and also because they strive to have their naval and air units consistently ready for combat at all times of the year. Spinney, on the other hand, verified that U.S. Navy carrier-based squadrons receive no combat training at all for the first two months after returning from a cruise, during which "flying operations are limited to maintenance check rides and instrument flight/airways navigation training."[98]

In the December 2001 issue of *Sea Power* magazine, moreover, Peterson found substantial differences between U.S. and British naval training programs, and the differences did not make the U.S. Navy look good. The article told the story of the lessons learned by the crew of the USS *Winston S. Churchill*, an *Arleigh Burke*-class destroyer, as she underwent Flag Officer Sea Training (FOST) Tier I training for two weeks in England. The Americans came away deeply respectful of the RN and its highly realistic training regime. Said one U.S. Navy sailor, "The Royal Navy brings realism to the next higher dimension.... The aircraft are flying below the bridge wing, the artificial smoke makes you gag, the voices on the 'comm' [communications] circuit are harried like they are under attack, and the OPFOR [opposing force] seems to come from nowhere—and that is just the Thursday War!"[99] Lt. Steven P. Murley, USN, remarked, "It's top notch. We've done a lot of things we don't do in the U.S. Navy—an opposed port breakout with aircraft attacking us in the breakwater, for example.... Our ORM [Operational Risk Management instruction] would not allow it."[100] Cdr. Guy W. Zanti, USN, said, "This training is outstanding.... I've witnessed many of the drills firsthand. The antiterrorism and force-protection drills were superb—no other ship in my 19 years of experience has had an antiterrorism and force-protection exercise to the depth and level that this crew received."[101] The American officers also recommended that more U.S. ships should undergo British FOST training, which is commendable, but one has to really wonder why the United States, the world's only naval superpower, has to rely on the now second-tier (at least in terms of hardware) Royal Navy for training. Is this what the world's greatest navy is supposed to do?

Realistic training is vital in maintaining combat readiness, and as demonstrated above, the U.S. Navy does not get enough of it, and for several reasons, not just ORM. For this reason, Capt. Daniel Appleton, USN (Ret.), has argued that U.S. Navy sailors "are not prepared to cope with the violence of battle."[102] More specifically, he suggests

> that many exercises are conducted hurriedly and with consequent reduced benefit. Since live exercises also rarely incorporate consideration of damage or human casualties, another result is an increased need for on-station battle drills to be adequately scheduled and skillfully planned and conducted. However, administrative and training paperwork plus maintenance workloads are now so heavy that battle and high-threat watch station team training is possible an average of only twenty minutes per week over a two-year training cycle—with more than 30 percent personnel turnover expectable during each year. Drastic means need to be developed and implemented to reduce administrative workloads on the crews of Naval ships in order to permit time to develop battle skills.[103]

12

Morale Issues, Racism, Drugs, Sabotage, and Related Matters

> The U.S. Navy is now confronted with pressures... which, if not controlled, will surely destroy its enviable tradition of discipline. Recent instances of sabotage, riot, willful disobedience of orders, and contempt for authority... are clear-cut symptoms of a dangerous deterioration of discipline.
>
> —HOUSE ARMED SERVICES COMMITTEE STATEMENT ON THE U.S. NAVY IN THE EARLY 1970s[1]

THERE ARE SEVERAL OTHER FACTORS that must be considered in evaluating the readiness of the U.S. Navy since World War II. Although most American sailors are decent and hardworking, over the years a significant number of them have proven to be not only unreliable but actually detrimental to combat readiness. During the Vietnam era, the combination of institutionalized racism against black sailors, a long and unpopular war, the draft, and an overworked fleet contributed to serious morale problems, violence aboard ships, disruptions of operations, mutiny, and sabotage. In 1971 "the Navy reported almost 500 cases of arson, sabotage, or willful destruction on its ships, while 1000 sailors on the USS *Coral Sea* petitioned Congress to stop its cruise to Vietnam. These 'flattop revolts' expanded the next year, as sailors signed petitions or disrupted operations on the *Kitty Hawk*, *Oriskany*, *Ticonderoga*, *America*, and *Enterprise*. Sabotage on the *Ranger* and *Forrestal* prevented their scheduled port departures while pilots became increasingly concerned about their role in the bombing campaign and questioned the war openly."[2] The USS *Ranger*, one of the mightiest warships in the world at the time, was taken out of action for more than three months, and it had taken only a single disgruntled U.S. Navy sailor (who was later acquitted) to do it.[3] Suspected sabotage was also detected on nuclear submarines in the 1950s, 1960s, and 1970s, including onboard the famous USS *Nautilus* (apparently on more than once occasion) and the USS *Spadefish*, again more than once.[4]

These issues are strongly related to an overall lack of motivation. It is safe to say that most sailors in Western navies join because their respective navies offer job security, training, and opportunities for travel and advancement. This is true in the U.S. Navy also, but there was one major difference between U.S. Navy sailors and other professional navy men, at least during the 1960s and early 1970s. As Freeman pointed out in his book about the disastrous fire that broke out on the USS *Forrestal* in 1967, most of the young sailors on the ship went into the Navy simply because they did not want to get drafted into the Army and sent to Vietnam.[5] Many of these young American sailors hated the military and never wanted to join the Navy but felt compelled to do so to avoid more dangerous duty elsewhere. Like the National Guard and reserve forces, the U.S. Navy was considered a safe and legal choice for those who wanted to avoid direct combat. In effect, they were, in a manner of speaking, voluntarily "conscripted" into the Navy so that they would not be involuntarily conscripted into the Army. This provides some additional context for understanding why the U.S. Navy had so many personnel problems in the Vietnam era.

Every navy has malcontents and disciplinary problems, but the U.S. Navy had much more than its usual share. Indeed, during the 1970s, Spector says, American sailors coming ashore for liberty in Greece appeared to be very undisciplined, leading one American to say that among U.S. sailors, "there is a lack of the tight discipline which seems to be apparent in other navies."[6] (According to Admiral Gorshkov, Kenyans came to the same conclusion in 1969, and he noted that Soviet sailors on shore leave were apparently far more disciplined and well behaved than their American counterparts.[7] Of course, one must also note that bad behavior ashore is not always an indication of poor discipline aboard ship, where it really counts. In this regard, many Western navies are also guilty.) For these unmotivated U.S. Navy sailors, sabotage and protests were just a means of registering their dissatisfaction with the draft and all things military. This is the price a democracy pays when its foreign policy starts taking on seemingly imperialistic undertakings in distant lands.

Making things even worse, during the Vietnam era the U.S. Navy did little to discourage senior, predominantly white, officers from really and truly living the good life while the enlisted men they commanded, who often belonged to minority groups, toiled in unnecessarily overcrowded conditions below decks. In the case of one carrier task force commander, the admiral's upscale cabin was described as being "like a New York Central Park South luxury apartment.... The Admiral's dining room... could seat ten comfortably around an oval table covered with starched white linen. The silverware was heavy and glistening. The meals were served with painstaking etiquette by white-coated attendants."[8] While the admiral lived in relative opulence and dined luxuriously, "belowdecks, the crew was jammed together, 150 men to each open windowless, poorly lighted, ill-ventilated bay. They lived one atop the other, three bunks high, with no privacy and little storage space, with the constant noise of the ship's operations jarring them. They took their meals in windowless, low-ceilinged mess spaces that doubled as warehouses for the bombs and rockets the airplanes would use."[9]

Naturally, disparities between senior officers and enlisted men are found in *every* fighting organization, but the problem is especially severe in the larger ships of U.S. Navy. This is because larger ships tend to be more socially stratified and less cohesive then smaller craft, and U.S. carriers are by far the largest warships in world. In addition, as Spector told us, unlike some other cultures, Americans tend to be less class conscious and do not easily accept that someone is "superior" just because he or she has a higher rank. British sailors, for example, have generally had less comfortable ships and have many of the same complaints, but "where the American sailors—and the American public—emphatically differed was that they refused to tolerate the notion that officers were in any way superior to enlisted men, save in rank and responsibility."[10] Thus perhaps there are luxuries that a British admiral may have that might not be advisable for an American admiral to have.

Sabotage and protests by American sailors have been powerful and recurring guest characters in the ongoing story of the U.S. Navy, but racism, as the late Adm. Elmo R. Zumwalt, USN, posited, was an "integral part of the Navy tradition"[11] until the early 1970s, and perhaps even later. In 1972, Lehman noted, on the USS *Saratoga* "no white officer would walk unescorted on the second deck, where the enlisted mess was. There were many incidents of racially inspired muggings and beatings by both blacks and whites and including some officers."[12] Certainly the institutionalized racism that rocked the fleet in the early 1970s is thankfully no longer so common, blatant, or obvious, but even some recently retired white senior officers are not convinced that racism is no longer a big problem in the U.S. Navy. Captain Abrashoff said in 2002, "Perhaps the most malignant obstacle to forming a cohesive unit is also the U.S. military's worst-kept secret: its inability to end racial and gender discrimination. Contrary to Pentagon hopes and hype, racism persists and sexual harassment is pandemic in nearly every military unit, land, sea or air. In fact, this shouldn't be surprising. The military, like any other organization, reflects the larger culture of which is it a part."[13]

Morale suffers for other reasons as well. In the past eight years, for example, there has been compelling evidence of serious morale problems among Navy junior officers. "In the fall of 1999," reported Jack Spencer of the Heritage Foundation, "the Navy surveyed its junior officers to gauge morale. They expected a 15 percent response rate, but, to their surprise, over 55 percent of those surveyed responded. Of these responses, 82 percent responded negatively. Citing poor leadership, inadequate pay and compensation, and insufficient spare parts and equipment, only one-third said they planned to reenlist."[14] Notice that the primary reason listed for low morale is "poor leadership," which, one might suspect, is a nice way of saying "bad senior officers and bad politicians, in that order." Overmanning and long deployments are probably also a factor in this equation.

Another problem is the U.S. Navy's low educational standards for enlisted sailors. In the 1950s, for example, approximately half of the sailors completing basic training were in the lowest intelligence categories, says Michael Isenberg, and "many of them could read at only a fifth-grade level."[15] In the dark days of the early 1970s, the Navy was forced to accept a large number of "Category IV" recruits, the least intelligent people who are allowed to serve. "In fiscal year 1971, 14 percent of new recruits were classified as in Group 4, while in fiscal year 1972, 20 percent fell in this group."[16] In 1977, "30 percent of all Navy recruits read below the 9th Grade level, although the majority were high school graduates."[17] Former Navy secretary Lehman submitted that in the late 1970s the U.S. Navy enrolled recruits "who were illiterate, convicted felons, drug users, and worse."[18] In 1985, Gabriel wrote that "the quality of personnel tends to be low" in the U.S. Navy and that many critical and highly technical positions could not be filled because of a shortage of well-educated sailors.[19] To compensate, the Navy hired civilian CETS (Contractual Engineering Technical Service) people to handle these duties aboard ships. Distressingly, the Navy became dependent on these civilians, who were not under any legal or contractual obligation to stay when the ships deployed. The commanding officer of an aircraft carrier lamented that he could not take even his ship to sea unless he had civilian contractors aboard to maintain some of the combat systems.[20] This too is a serious weakness, although it may not be unique to the U.S. Navy.

The admirals gloried about their "high quality" all-volunteer force in the late 1980s and early 1990s, but in 1993 it was reported that the Navy still had thousands of sailors who could not read material designed for a junior high school audience. The Navy confessed that despite its efforts to attract high-quality people, "a quarter of their recruits can't handle manuals geared to a ninth grade reading level."[21] These days, the required AFQT (Armed Forces Qualification Test) scores required to join the Navy are still relatively low. According to Rod Powers, the minimum acceptable score for the Navy is just 35 (and prior to 2003 even lower), whereas the Coast Guard and Air Force each requires scores of at least 40. The Navy will also accept 5 to 10 percent of recruits who are high school dropouts with GEDs, as long as their AFQT scores are higher than 50. The U.S. Air Force is far more selective and does a relatively good job at compensating for the unreliable U.S. education system, in that it rarely admits anyone who does not have a high school diploma, and even these folks must have an AFQT score of 65 or greater.[22]

Even though the vast majority of American sailors today are high school graduates, it must not be forgotten that due to low standards in the U.S. public school system, "American high school graduates are among the least intellectually competent in the industrialized world."[23] Correspondingly, and predictably, "Americans are at or near the bottom in most international surveys measuring educational achievement,"[24] especially in math and science. American universities, with some very notable exceptions, are not much better than the high schools as regards producing intellectually competent graduates. With the exception of certain Ivy League schools (but not all of them) and a few others, American universities are generally not very selective. Almost anyone in the United States with a high school diploma (and the money to pay for tuition) can gain admission to a college or a university, because many U.S. institutions of higher learning actually have very low standards. "A College Board survey of 2,600 colleges showed that only forty percent required any minimum grade point average for admission and only thirty percent set minimum cut-off scores on the SAT," says Anelauskas.[25] And while fewer than

10 percent of U.S. high school graduates will get accepted into Harvard, Anelauskas comments, European schools are even more demanding. "According to literacy studies by the National Assessment of Educational Progress, fewer than five percent of American high school graduates would meet the entrance standards of European universities."[26]

Even some people who *graduate* from U.S. colleges have very poor reading and critical reasoning skills. (Apparently at least a few of them have managed to become officers in the U.S. Navy. Karam notes that one of the junior officers on his ship was "barely literate" and needed his help writing reports.) This crisis in education was illustrated well in recent studies by researchers at the Educational Testing Service. According to one of the researchers, "the literacy levels of U.S. college graduates 'range from a lot less than impressive, to mediocre, to near alarming.'"[27] As Anelauskas surmises, "Surely if illiterate persons can graduate from American universities, it is not far-fetched to say that something is uniquely wrong with the American education system."[28]

According to the *Times Higher Education Supplement* of November 5, 2004, thirty-five of the hundred top rated universities in the world are in the United States, but as Professor David VandeLinde of Bath University has said, there are tremendous variations in quality throughout the U.S. system, and we must temper this statement with the likelihood that the world's hundred *worst* universities are probably also American.[29] Grade inflation and easy classes are to be found in almost every major American university these days. Even U.S. doctoral programs, long considered the "gold standard" of education, are sometimes of poor quality; as sociologist Dr. Barbara Lovitz has declared, "Evidence suggests that poor quality dissertations are often passed. Adams and White (1994), in a study that looked at dissertations abstracts, found that a significant number of dissertations that had passed had obvious and sometimes fatal flaws."[30] Dr. Yoon Tae-Hee, president of Seoul University of Foreign Studies and an adjunct professor at Clemson University, South Carolina, said in 2005 that "anyone who is not mentally retarded can get a Ph.D. in the United States."[31]

The Naval Academy itself is also far from innocent in these matters. In 2002, Cdr. Gerald L. Atkinson, USN (Ret.), indited, "There is documentary proof of lowered academic standards at the Naval Academy. In 1990, a civilian chairman of the electrical engineering department was relieved of his post in mid-semester because he refused to raise preliminary grades in two electrical engineering courses and refused to raise grading curves 'across the entire [electrical engineering curriculum].'"[32] Atkinson declared, "It is clear that the U.S. Naval Academy has been slowly and subtly but determinedly lowering standards over time at the Navy's premiere source of naval officers."[33] In a 1996 interview with Annapolis professors, he noted, "they explained that 'fully 30 percent of the midshipmen in their classes were not qualified to be in any college, much less the Naval Academy.'"[34] Finally, "In the early 1990s, the Academy added a Counseling Center, remedial courses, outside contracting for teaching remedial reading and writing. . . . Courses that have been identified as too challenging have been eliminated or 'redesigned to be more reasonable to the needs of today's midshipmen.'"[35] The Naval Academy is still considered to be one of the most selective universities in the United States and one of the most distinguished.

In sharp contrast are the sailors and officers of today's Japanese Maritime Self-Defense Force. Japanese sailors are universally and highly literate[36] and exceptionally well educated. Japanese schools are notoriously tough academically; "experts believe that an average high school education in Japan can be equated with an average college education in the United States."[37] It is very rare indeed to see overweight Japanese sailors, and even now the drug abuse rate in Japan is "still very low compared to that in other counties."[38] Drugs are probably not a big problem in the Chinese People's Liberation Army Navy, either, for drug traffickers in that country can be and have been put to death.

On the other hand, drugs have most definitely undermined the combat readiness of the U.S. Navy, and the problem was especially noticeable in the 1970s and 1980s. The drug situation aboard ships in the 1970s was described well by Spector:

> Cheap and plentiful supplies of drugs were available to sailors when their ships visited Subic Bay and in many Mediterranean ports. Aboard many ships there was an elaborate substructure for the acquisition, concealment, sale, and distribution of drugs. At the top of the underground structure was 'the boss or head pusher,' most likely a 'petty officer of E-4 to E-6 level. He is in business for money.' The head pusher presided over a network of drug runners, addicts, and habitual and casual users. . . . One senior officer noted that at least 'a few of the more stable and intelligent experimenters and moderate users [may] have become senior petty officers,' and that abuse among commissioned officers was far from unknown.[39]

In 1981, it was revealed that 15 percent of the crew of the submarine USS *Parche*, including three of her officers, failed a drug test just before a scheduled deployment. The *Parche* was used for spying operations, and many of these covert visits to Soviet waters were extremely dangerous. Sontag et al. explain that some American submariners turned to drugs to help deal with the stress of dangerous intelligence gathering deployments or to avoid going to sea altogether. "*Parche* wasn't unique in her personnel problems, and the drug bust had intelligence officials worried," they adduced. "*Seawolf*'s crew was disintegrating under the mounting frustrations of serving on a broken-down and cursed boat. The pressure inspired some of her crew to lose themselves in a marijuana haze. *Some even proclaimed their drug use openly and loudly, just to get off of* Seawolf"[40] (emphasis mine). Karam also reports that more than one U.S. Navy nuclear powered ship has been "shut down from time to time because of sloppiness or in one case (a submarine), for excessive drug use by the crew."[41]

None of this should be really surprising to anyone who has paid attention to the U.S. Navy for the past twenty-five years, except possibly Tom Clancy. Americans frequently deride the Soviet/Russian armed forces for having serious substance abuse problems, but all too often they overlook the severity of the problem in their own armed forces. It seems to some that this is really just another case of "the pot calling the kettle 'black.'" In the early 1980s, for example, "Drug and alcohol abuse was rampant throughout the fleet," writes Gregory Vistica. "Forty-seven percent of the Navy's personnel was smoking marijuana. Another 11 percent was snorting cocaine."[42] A 1981 crash on the deck of the USS *Nimitz* killed fourteen sailors, and half of their bodies contained traces of marijuana. The Navy introduced mandatory random drug testing to counter this problem, but random testing by urinalysis definitely has its limitations. In the last six years, naval aviators, SEALs, and other officers have been arrested for drug trafficking or usage, along with at least ten thousand enlisted personnel. In 2001, for example, the U.S. Navy discharged more than 3,400 personnel for drug abuse, which is "more people . . . than the Army, Air Force and Marine Corps [discharged] combined."[43] Not even the great bastion of American navalism, the U.S. Naval Academy, is immune to the scourge of drugs. In his famous 1971 article "The Collapse of the Armed Forces," Col. Robert D. Heinl Jr., USMC, wrote that drug use at the Naval Academy was "anything but unknown,"[44] and almost thirty years later Burns lamented, "It used to really mean something to be a Naval Academy graduate. In recent years they've had pedophiles, car theft rings, drug rings, cheating scandals and murderers."[45] This precipitous decline in the moral and ethical quality of a substantial number of Annapolis midshipmen would tend to explain why Lt. Donald Johnson, USN (Ret.), devoted an entire chapter of his rather unpopular 2002 book *It Wasn't Just a Job: It Was an Adventure* to the "Immoral and/or Unethical Acts" he had witnessed during his twenty years in the Navy.[46]

The Navy's own figures indicate the drug and alcohol problem has been reduced substantially since the 1980s, but the evidence supporting that claim is quite unreliable, and in a warship disaster can result if even a few sailors are high while under way. Why should one be skeptical of the Navy's claims of "winning" the drug/alcohol war in its own ranks? Well, it should be pointed out that certain designer drugs now available pass through the body very quickly and are more difficult to detect than others by urinalysis. In addition, the drugs of choice have changed since the 1980s. As the American Civil Liberties Union reported in 2002, "Because urine testing is based on an analysis of metabolites associated with the drug in question, and because alcohol and cocaine, heroin and methamphetamine all pass through the body within 24–48 hours, leaving no metabolites, marijuana is the only substance that is easily detected with urinalysis. Drugs that have a more significant impact on employment or work performance, such as alcohol and other illegal drugs, are not effectively tested for with urinalysis."[47] Some have argued that mandatory drug testing has actually prompted sailors to drink more or to use more dangerous drugs than marijuana. Jacob Sullum said in 2003, "Thus to the extent that drug testing encourages servicemen to drink rather than smoke pot, it may actually impair effectiveness. A former Navy psychiatrist reported another perverse effect of the military's 'zero tolerance' policy: He said a member of an aircraft carrier's crew told him LSD was more popular at sea than marijuana because it was much easier to conceal. 'Instead of occasionally smoking a joint on a carrier,' the psychiatrist said, 'guys were tripping! Oh, my God. That scared me.'"[48] And in his 2004 book, Doug Thorburn referenced a study from the late 1980s that indicated "an astonishing 46% of recruits 'with an identifiable history of problem drinking' at the time of enlistment" in the U.S. Navy.[49] This, along with the evidence uncovered by a 1994 National Academy of Sciences report that random drug testing actually has little or no deterrent effect,[50] makes it easy to speculate that the number of *actual* drug and alcohol abusers in the U.S. Navy could be much greater than the number who are caught these days. Other navies certainly have drug and drinking problems; however, no navy in the world is more closely associated with the drug problem than the U.S. Navy. When one acknowledges that the U.S. Navy has *knowingly* accepted felons and drug users to fill its ranks, that American teenagers have the highest "alcohol-and-drug-abuse rate of any industrial nation,"[51] including liberal countries like the Netherlands, then it appears that drug abuse in the U.S. Navy is just another manifestation of a massive American criminal subculture. Drug and alcohol abuse is undeniably self-destructive behavior, and even Congress has proclaimed that America is "the most violent and self-destructive nation on Earth."[52] For these reasons, some remain unconvinced that the U.S. Navy no longer has a serious drug/alcohol problem.

Poor physical fitness is also evident in the ranks of the U.S. Navy. In a 2001 column, investigative journalist Jon Dougherty revealed that a survey conducted by a team of doctors at Marywood University had found that many personnel in most branches of the U.S. military were unfit, overweight, and physically inactive. Abdominous U.S. military personnel were almost as unfit and inactive as American civilians, in fact. The survey reckoned, "Military personnel do not exercise any more than the general population, even though some amount of physical training is required in all branches."[53] This is unsettling news to be sure, especially for Americans, who on average are among the most overweight people in the world. (According to the OECD, in 1999 some 22 percent of Americans were obese, versus about 20 percent in England, about 10 percent in France, Denmark, Sweden, Italy, and Norway, and 3 percent in Japan and Korea.)[54] Dougherty, who served as a corpsman in the U.S. Naval Reserve, found a simple but valid explanation for this embarrassing situation: "Our overweight, undertrained, physically unfit military is little more than a reflection of American society as a whole, I fear."[55] In 2005, a U.S. Army nutritionist warned that the increasingly mastadonic average American is "quickly becoming a national security issue for us."[56] As the sage old saying goes, "The apple does not fall far from the tree."

To put it as delicately as possible, and with all due respect, there is nothing in the U.S. Navy that does not exist in American society, and that includes a substantial and lingering historical legacy of racism, substandard public education, widespread obesity, and drug and alcohol abuse.

13

What Tom Clancy Does Not Know or Won't Tell You

> We are inclined to overestimate our ability and underestimate our vulnerability.
>
> —Cdr. Dale Sykora, USN, former skipper of the nuclear submarine USS *Dallas*, 2004[1]

Through his many best-selling books and movies, author Tom Clancy has created a crisp, sharp, spit-polished, efficient, and patriotic image for the U.S. Navy. Some think he might be a paid public relations consultant or recruiter for the American submarine force. It may come as a shock to some of his readers, however, that the American ships, submarines, aircraft, equipment, and sailors in his books are too good to be true. In 2001, Shuger suggested that Americans have placed too much stock in Clancy's writings, and that is perhaps especially damaging since Clancy has moved from novels to nonfiction. The result, Shuger worries, is that "millions and millions of people . . . have gotten most of what they know about warfare and the U.S. military from an ex–insurance agent who never served a day on active duty."[2]

Furthermore,

> Does he know what he's talking about? He certainly seems to know a lot about how planes, subs, and missiles are supposed to work, and how we and the Soviets intend[ed] to use them. And this makes his books that much more seductive. But is there any reason to think that he knows

what will happen when those weapons and those intentions are put into the pressure-cooker of combat? The more complex war has become, the more ways there are for missions to go bad, and the graver the consequences. The history of modern warfare is replete with counterexamples to Tom Clancy's vision. The problem is that history hasn't sold twenty million copies. How unlike fiction is real war! Clancy has it in his head—and his readers are getting it drummed into theirs—that the U.S. military is a precise instrument, capable of almost effortless accuracy.[3]

Luckily, however, Clancy has competition, and what is more, the competition is much more candid, credible, authentic, and realistic. Several recent books have effectively stripped off much of the shiny Hollywood polish on the American submarine force, most notably former Petty Officer (now Dr.) Andy Karam's account of life on the USS *Plunger*, titled *Rig Ship for Ultra Quiet* (2002), and Douglas C. Waller's *Big Red* (2001). Both authors made it known that though there is a lot of hype regarding U.S. submarine training, the reality is much less impressive. As for the legendary assertion that all U.S. submariners are experts on "every system" in their boats, one sailor told Waller that was "all bunk." Waller explains, "The [submariner's] qualification only made you *familiar* with the rest of the boat. It didn't mean you could actually run other parts. If [the sailor] and the other missile techs suddenly died, those nukes in the back wouldn't have a clue how to fire these rockets."[4] Karam, an engineering laboratory technician who eventually became a chief petty officer in the Naval Reserve, concurs, acknowledging that he could work on other systems only "in a pinch."[5] He continues, "The *Plunger*, and, for that matter, any nuke boat, was sufficiently complex that one person simply could not learn everything to that level of detail in the 14 months we were given to qualify. Not if they were doing their own jobs, too."[6] One will not find this awful kind of truth in any Tom Clancy book. The nonfiction he produces on submarines is well written and detailed, but it is still essentially, at its core, Navy PR fluff.

British allies, of course, have long questioned why American submariners spend so much time and effort on nuclear reactors. Surprisingly, Waller has written, some U.S. Navy officers quietly agree. The drill coordinator on the USS *Nebraska*, Lt. Brent Kinman, USN, told Waller that American submariners talk too much about the reactor, like mechanics, and not enough about how to fight the ship effectively: "That was the problem with today's submariners, Kinman thought. They were technicians rather than warriors. The average lieutenant riding these boats considered himself a nuclear engineer first and a submarine officer second. 'It almost feels like we're out there just driving the reactor around.'"[7] Said Spector, "Critics also charged that nucs neglected seamanship and navigation. A report by Commander, Submarines, Atlantic Fleet, in 1979 cited fourteen major submarine accidents that were, in whole or in part, due to 'less than sufficient performance with respect to seamanship.' A retired nuclear submarine skipper declared than 'an American submarine might run aground due to total incompetence in navigation and ship handling, but the reactor-control division records would be perfect as it hit.'"[8] This overemphasis on engineering might explain why diesel submarines so often triumph against American nuclear submarines during exercises.

In his controversial, "loved or hated" 1986 book *Running Critical*, Patrick Tyler presented evidence to suggest that the mainstay of the U.S. nuclear submarine force, the first of the *Los Angeles* (SSN 688) class, was not a first-class attack submarine and that, like the F-4 and the F-14 fighters, it was possibly more expensive than combat effective against potential adversaries. According to Tyler, a former CNO, Admiral Zumwalt, was not impressed with these submarines: "To Zumwalt, the *688* submarines were a misbegotten class: shallow-running, unstable in tight turns, vulnerable at high speed, and too costly for the marginal advance they had given the navy over previous alternatives."[9] U.S. Navy crews complained that the boats were incondite, built by reportedly lazy and indifferent shipyard workers, and based on a compromised committee-driven design that was demonstrably inferior to Soviet contemporaries in all areas except stealth, and that even in that parameter they were inferior to diesel submarine contemporaries, of which the Soviets and many others had plenty. The result was a thoroughly mediocre submarine. Karam also thought the original *Los Angeles*-class boats were less than stellar performers, especially when they were deployed on "spec-ops" (spying). "At the time I was in, *LA*-class subs were fairly routinely detected on spec-ops—it seems they had a tendency to lose depth control at periscope depth because their fairwater planes were too high on the sail. I understand the improved *688*s are better. The USS *New York City* was apparently detected routinely on one mission every time they streamed their floating wire antenna because sea gulls sat on the wire and hitched a free ride."[10] Even Admiral Rickover once said, in the early 1970s, when the U.S. Navy was in particularly bad shape, that if he had the choice of commanding either the Soviet or the U.S. submarine fleet in war, he would prefer the Soviet one.[11]

Of course, American nuclear submarines have successfully attacked allied surface ships and diesel submarines on exercises too, and it would be unfair and remiss of me not to mention that, but nevertheless, many allied and NATO officers are not overly impressed by American nuclear submarines or their crews. Compton-Hall, for example, lavishes all but unqualified praise on the Dutch, Canadian, German, Australian, and Scandinavian submarine services, but his comments on the American silent service are decidedly mixed.[12] Praise was included, but it was infrequent and substantively qualified. For example, like Shuger, he says that the U.S. Navy submariners are superb engineers, "but there is a case for saying that fighting capabilities took second place over a long period during the Rickover reign."[13] Sprinkled through his discussion on the Americans are terms such as "Rickoverized" (which means obsessed with engineering), "dogmatic," "conformism," "conservation," and "complacency." He also takes issue with the U.S. Navy's "habit of overstating fitness reports which is no kindness to the man or to the service. The effects have been felt far beyond deserved or undeserved promotions. A serious result has been that the cold-blooded, highly critical post-attack autopsies of the kind to which British command teams are traditionally, and often embarrassingly, subjected have generally been avoided. Avoidance has led to over-confidence and lessons not being learned."[14] Shuger agrees that Navy officer fitness reports are often far from candid. "Almost everyone who hasn't been court-martialed gets mostly A's in the set categories. And the narrative material is supplied by the officer himself and then mega-hyped by his immediate superiors into a superlative-soup that renders distinctions difficult."[15]

This same oblivious behavior was noted by a former captain of the Canadian pocket carrier HMCS *Bonaventure*, who reminisced back to the year 1968: "I do remember an American nuclear submarine getting his comeuppance when he was attacked by one of our Trackers. They thought they were absolutely foolproof. This guy was discovered, pinged on and attacked and nailed, the whole schmeer, and he couldn't figure out why. We weren't anxious to tell him either."[16] He concluded by saying that the defeated U.S. submariners were amazingly and unjustifiably "overconfident."[17] And in the joint Indian Navy–U.S. Navy exercise MALABAREX in 2003, the frigate INS *Brahamaputra* took on the seemingly formidable *Los Angeles*-class nuclear submarine USS *Pasadena*. The Indian ship "was able to detect *Pasadena* 'from over 8 miles away,' and engage it, 'getting a mission kill' in the process," said the Indian magazine *Frontline*.[18] It would be reasonable to assume that the crew of that sub was also surprised by the result, although the hydroacoustical conditions in the exercise area might have been unfavorable to the submarine.

One last thing that you will not find in any Tom Clancy book is any substantive discussion on the U.S. Navy's safety record since the loss of the submarines *Thresher* and the *Scorpion*. During the past fifty years or so, it was quite fashionable for U.S. sailors to mock the Soviets, especially as regards safety. Soviet nuclear submarine reactors were often poorly designed, not as safe as their American equivalents, and I do not think anyone would dispute that. What some forget, or perhaps discount, is that the U.S. Navy's safety record has also frequently been called into question.

In both 1989 and 2000, the U.S. Navy had to order all units to stand down from normal operations to review basic safety procedures, in both cases after a string of serious accidents.[19] The same thing happened once again (in naval aviation only) in 2006 as well. Everyone familiar with the U.S. Navy knows about the losses of the *Thresher* and the *Scorpion* in the 1960s, which forced the U.S. Navy to improve its submarine safety programs, but there have been other, one might say, "near misses" since then that have not gotten quite so much attention. "While routine U.S. policy is to deny the possibility of nuclear reactor accidents," noted Hayes et al., "the fact remains that many nuclear warships *have* experienced mishaps."[20] According to Arkin and Handler, in 1973 the USS *Greenling* sank below her test depth for a short time, simply because one of her depth gauges malfunctioned.[21] Such a descent can be fatal, needless to say, but luckily this time it was not. In the following month, the USS *Guardfish* "experiences a primary coolant leak while running submerged.... The submarine surfaces and is ventilated and decontaminated, and repairs the casualty unassisted. Four crewmen transferred to the Puget Sound Naval Hospital for monitoring."[22] This was a serious accident, for as Shay Cullen put it, "Without this vital fluid [the coolant] the reactor will overheat and melt down, the worst nuclear disaster possible." The crew "barely managed to avoid a meltdown. There was a serious radioactivity leak but what was more serious was the cover up, the deck log book and the command history were falsified."[23]

In a situation like this, in which there is loss of coolant, the repercussions can be very serious. Said DiMercurio:

> The fuel in a nuclear reactor is 'bomb grade,' unlike a civilian reactor; it uses U-235, the high-octane variety, instead of natural uranium.... If a

U-235 bomb-grade uranium core melts in a loss of coolant accident, there is a possibility that it could form a *critical* mass at the bottom of the core. It is further possible that an uncontrolled nuclear reaction would then take place. In the least likely case, it would explode like a fission bomb and vaporize the ship. In the more likely case, it would cause a 'prompt critical rapid disassembly,' which is an uncontrolled runaway reaction that causes the nuclear fuel to thud in less than a full explosion but that is strong enough to blow open the reactor vessel and the hull.[24]

This is not very reassuring, but there is more. In 1975, "The USS *Haddock* (SSN-621) develops a leak during a deep dive while on a test run near Hawaii. The U.S. Navy confirms the incident, but denies the vessel is unsafe as crew members had charged in late October. A number of enlisted men had protested sending the ship to sea, claiming it had cracks in the main cooling piping, leaks, and malfunctions and deficiencies in other systems, including the steering mechanism."[25] Fortunately, there was no loss of life. Interestingly, it should be noted that a near-fatal accident on the USS *La Jolla* in the 1981 was caused by none other than the father of the Nuclear Navy, Adm. Hyman G. Rickover, USN, himself. As Tyler tells it, thanks to Admiral Rickover, a bureaucratic tyrant par excellence, the ship went out of control and nearly reached her test depth; if she had continued her uncontrolled dive much longer, "that would have been it—another *Thresher*."[26] A few years later, DiMercurio, then a nuclear submarine officer, admitted that he "almost melted down a nuclear reactor," although he did not say where or how (but he did mention that he served on the nuclear submarines USS *Pargo* and USS *Hammerhead*).[27] In 2002, additionally, the research submarine USS *Dolphin* was also nearly lost because of a fire and flooding, and the crew had to abandon ship. And of course, there have been quite a few collisions (more than twenty, by some accounts) between U.S. and Soviet/Russian submarines, and while it would be unfair to place the blame entirely on the U.S. Navy, it is well known that "the gung-ho attitude of American sub commanders, who regularly ignored the rule that they should maintain a constant distance from their 'target' of at least five miles," was probably a salient factor.[28]

What does all this mean? It means, by all accounts, that the U.S. Navy has been lucky, *very* lucky, for as one U.S. submariner who served on the USS *Sargo* said in 1983, "I'm really surprised we only lost two subs. . . . There were times when we weren't sure we were coming back."[29] As I said before, every navy has accidents, but given that the U.S. Navy has had not one but three official safety stand-downs since 1989, mistaken an airliner for an F-14 and shot it down, accidentally fired a missile at a Turkish destroyer (killing her captain and several others), and in 1991 killed two Canadian divers through negligence on the USS *Pharris* (the Canadians were doing an inspection of the *Pharris* because, unlike the Canadian navy, U.S. destroyers don't usually have their own diving teams),[30] one might actually form the impression that the U.S. Navy is somewhat accident-prone and bunglesome, and its denials and cover-ups only serve to reinforce Shuger's 1996 statement that it is not a "reality-based" organization.[31]

If there is anything the U.S. Navy needs, it is a healthy dose of reality. Case in point, there was a friendly-fire accident involving the battleship USS *Iowa* and one of her escorts that every American should know about. The destroyer USS *William D. Porter* accidentally launched a torpedo at the *Iowa*, which just missed. The near miss was bad enough, but even worse, there were VIPs aboard the *Iowa* during that cruise, including the president of the United States and the Chief of Naval Operations! The captain of the *Iowa* wanted to court-martial the skipper of the destroyer but was overruled by Franklin Delano Roosevelt so as to avoid "adverse publicity."[32] Both presidents FDR and George H. W. Bush have in some way tried to gloss over the Navy's friendly-fire problems, and that too is scandalous. Thanks to presidential interference, the U.S. Navy was spared a much-needed reality check, one that might have taught it a lesson and perhaps spared lives in future accidents.

In addition to the aforementioned submarine accidents, friendly-fire incidents and near misses, there have been other accidents involving nuclear weapons. According to the Center for Defense Information, in 1959 "a U.S. Navy P-5M aircraft carrying an unarmed nuclear depth charge without its fissile core crashed into Puget Sound near Whidbey Island, Washington. The weapon was never recovered."[33] Six years later,

> [A]n A-4E Skyhawk strike aircraft carrying a nuclear weapon rolled off an elevator on the U.S. aircraft carrier Ticonderoga and fell into the sea. Because the bomb was lost at a depth of approximately 16,000 feet, Pentagon officials feared that intense water pressure could have caused the B-43 hydrogen bomb to explode. It is still unknown whether an explosion did occur. The pilot, aircraft, and weapon were lost. The Pentagon claimed that the bomb was lost "500 miles away from land." However, it was later revealed that the aircraft and nuclear weapon sank only miles from the Japanese island chain of Ryukyu.[34]

Although some doubt the weapon could explode simply because of water pressure (at that depth, roughly seven thousand pounds per square inch) or that if it did explode we would not know about it, the loss of a hydrogen bomb in itself is disquieting. Incredibly, or perhaps not so incredibly, the U.S. Navy did not inform the Japanese of this accident until the mid-1980s![35] Hayes et al. also report that the U.S. Navy acknowledged that some 381 "naval-nuclear accidents and incidents occurred between 1965 and 1977 on all types of vessels, especially involving anti-submarine and air-defense nuclear rockets."[36]

And indeed in 1992, a leaked U.S. Navy document ("OPNAVINST 3040.5B. Nuclear Reactor and Radiological Accidents: Procedures and Reporting Requirements for") gave the commanders of U.S. Navy ships visiting foreign ports the discretion *not* to inform the host nation in the event of a nuclear accident.[37] Perhaps this is why New Zealand still refuses to allow U.S. Navy ships to visit its ports, as it has since 1984. This poses a most disturbing dilemma for the United States should other countries follow New Zealand's lead. There certainly have been anti-nuclear warship/weapons protests in Australia, the Philippines, and Japan, for example, which all provide key facilities to the U.S. Navy. And, I must point out, with Hayes et al., "without their active support, U.S. capacity to fight an offensive nuclear war, to project power and especially to achieve 'maritime supremacy,' would be severely constrained."[38]

14

Misleading Congress, and a Cultural Explanation

> And you've got to consider the psychology of the Navy itself....
> The Navy, traditionally, technically, doesn't do anything wrong.
>
> —Former Army research director Raymond Walker
> on the Navy's flawed investigation of the
> USS *Iowa* explosion in 1989[1]

THE QUESTION THAT NOW REMAINS IS: How has the U.S. Navy managed to conceal all its glaring faults, bad policies, and appalling weaknesses for all these years? Part of the answer is that the Navy has a history of not telling the full truth to Congress. It is well known that senior U.S. Navy officers have a tradition of omitting information about the Navy's weaknesses and deficiencies during public testimony. For example, in the early 1980s, wrote Scammell, Navy officers tried to conceal the shortcomings of the new Aegis system by using unrealistically easy operational tests, then by classifying the poor results: "An amalgam of sophisticated seaborne radar, computers, and surface-to-air rockets ten years in development, Aegis was built to simultaneously track up to two hundred aerial targets and to control thirty killer missiles. But in sea tests against sixteen easy targets—easy because they were lobbed in one after another instead of all at the same time, as they would arrive in combat—the supershield missed all but five."[2] Consequently, "The results of the sea trials were immediately classified, ostensibly for reasons of national security, and it was announced that the tests had been successful. When Congressional overseers eventually learned they had been duped—a gain because not everyone in the fiasco interpreted 'patriotic duty' as 'staying silent'—the Aegis program was very nearly scuttled."[3] According to Representative Denny Smith, a Republican from Oregon and former F-4 fighter pilot, Navy officers deliberately deleted key passages from their initial test reports on the Aegis system to keep him in the dark on its failings.

This attempted cover-up was certainly not an isolated incident. Indeed, attempts by the Center for Naval Analyses to evaluate the Navy's way of doing things have been subjected to political pressure not to let on about any "implied weaknesses in current hardware or doctrine,"[4] said Shuger. And I suppose by now no one should be shocked to hear that in the mid 1990s, during tests of a new guided missile, the missile "melted its on-board guidance system. 'Incredibly,' an Army review said, 'the Navy ruled the test a success.'"[5]

Another part of the answer is that building ships, submarines, and aircraft for the U.S. Navy is big business, and with a few salient exceptions like Smith, pork-barreling throttlebottom politicians may not want to hear that the systems being built in their districts (and providing many, many jobs to voters) won't work or are not really needed. When Representative Smith tried to hold the Navy accountable for the botched tests of the Aegis system and the attempted cover-up, "Trent Lott, the House Republican whip, asked Smith if he knew that killing the Navy's Aegis cruiser program could affect sixteen thousand jobs at Ingalls Shipyard in Lott's home state of Mississippi."[6] Lott is well known for his devotion to the shipyard in question, where his father once labored. He even forced the Navy to buy an additional ship that it did not request so as to keep jobs at the facility: "I'll do anything for that [Ingalls] shipyard."[7] On top of that, noted Holland,

> The case of the Navy's F/A-18 [Hornet] strike fighter program also demonstrates that even legislators generally critical of the military will act to protect a program in which constituents' jobs are at stake. When President Carter threatened to cut the Hornet program in 1978, Representative Tip O'Neil (D-Mass.) and Senators Ted Kennedy (D-Mass.), Alan Cranston (D-Calif.), and Thomas Eagleton (D-Mo.) intervened. The planes are constructed by Northrop (California) and McDonnell-Douglas (Missouri), and the engines are built by General Electric (Massachusetts). In 1972, when Senators Lloyd Bentson (D-Tex.) and Thomas McIntyre (D-N.H.) led resistance to Defense Secretary Melvin Laird's request for

production money for the Trident submarine until the military produced an approved design, Senator Henry Jackson (D-Wash.) organized congressional support and was rewarded with a Trident base in his state."[8]

With politicians who are much more interested in jobs than effective and properly tested weapons, and officers who do not always like to reveal the truth (of course, as we have seen here, some of today's naval officers, especially the reformers, are more candid about the Navy's deficiencies), the U.S. Navy has great difficulty maintaining its credibility both in government circles and at sea.

There is one final possibility that comes into play, and it is cultural, deeply entrenched, and difficult to remove. My maternal grandparents were American, and as a child I went to public schools in the United States. I have also had experience as a student or as a lecturer in two other countries. Accordingly, I have a reasonably sound basis for making comparisons, both educational and cultural. It has been my observation (and that of many others) that American public schools and culture place much more emphasis on cultivating self-esteem than do those of other countries. This has led to a false confidence, or bravado if you will—a false sense of self-importance and, to a certain extent, plain, old-fashioned narcissistic egotism. As a matter of fact, this streak of overconfidence goes back a long way in American history. Samuel Huntington, in his erudite book *The Soldier and the State*, made a scorching reference to it during the time of the great American navalist Alfred Thayer Mahan: "Not only were Americans bellicose, but they suffered from an overweening and highly dangerous self-confidence. The military officers expressed great alarm at the 'national conceit' rampant in the United States."[9] This overconfidence is manifested today in several ways. It manifests itself in international mathematics competitions in which American teenagers get the highest scores in "self-confidence" but the lowest scores on the actual exams.[10] It is manifested when young Americans, with absolutely no musical talent whatsoever, audition on TV shows like *American Idol* and then break into tears when a judge tells them something ridiculously obvious, like "You're just awful. You have no talent. This was a waste of my time"—things they should rightly have known long beforehand. And it is also manifested in naval exercises, where American units are sometimes shocked that a competent enemy with good tactics can easily defeat them. Sadly, this overconfidence, combined with a misinformed and pork-loving Congress, may someday have profound consequences for the U.S. Navy.

15

Conclusion

THE U.S. NAVY IS THE LARGEST NAVY in the world, and on paper certainly the most powerful. Many believe it is the best navy the world has ever seen, and it is also unmistakably the most expensive. On the latter point at least, there is no doubt. With the Russian navy all but gone and the Chinese People's Liberation Army Navy still ascending, the American navy remains the largest sea power in the world. Yet, as we have seen here, this heavyweight navy often has great difficulty handling the little guys. Indeed, if the U.S. Navy were a boxer, one might say that his dominance is due mostly to his sheer size, because he punches well below his massive weight. In this era of asymmetrical warfare, of David-versus-Goliath conflicts, perhaps it is time for America to rethink its naval strategy, lose some weight, and, as sports announcers say, "focus more on the fundamentals." For all the money America spends on its huge navy, it really needs to be much better. To achieve this, I recommend the following simple twelve-step program:

1. Discontinue the "up or out" promotion system and use the systems used by all other English-speaking countries.
2. Start designing and building non-nuclear AIP submarines.
3. Restore the MCM forces to a level proportionate for a great power, so that there will be no need to depend on anyone else for that mission.
4. Ensure that all weapons systems are thoroughly, independently, and realistically tested.
5. Recruit better educated people, like the U.S. Air Force does. Also, said former sailor Steve Cook, "Give enlisted sailors a better

quality of life—food, quarters, uniforms, and overall fair treatment, especially for the low-ranking enlisted sailors, so that they feel like part of the team, instead of feeling like the lowest of the low. Quit making 'chief' the focus of everything, and engage the younger troops. The good ones are eager to make a useful contribution and want to learn how to become better sailors. Maybe the Navy should assign mid-ranking petty officers (E-5–E-7) as mentors for the younger sailors."

6. Follow Admiral Rickover's recommendation and reduce crew turnover.
7. Cease being so arrogant and learn from the best practices of other navies (this also means less engineering and more tactical training for officers).
8. Seek alternatives to the big-carrier-centered fleet, as recommended by Admiral Turner, and build more surface escorts, as recommended by Admiral Zumwalt.
9. Improve standards of physical fitness.
10. Petition the Congress to eliminate the joint duty requirement for officers.
11. Study and openly discuss the problems with excercises and ensure that all exercises are monitored by neutral observers who cannot be pressured to overlook failures.
12. Aggressively pursue and eliminate racism, sexism, and drug and alcohol abuse in the ranks by recruiting a higher caliber of sailor. Be more selective in recruiting.

Edmund Burke once said, "A nation without the means of reform is without the means of survival."[1] So too, I would add, is a navy. We should also heed the words of Lt. David Adams, USN, who said in 1997, "We must not forget that a determined adversary can shatter our illusion of invincibility and turn our confidence in low-cost military victory into devastating political defeat."[2] Finally, as a friend, I take my leave by asking the great American public to draw inspiration from the reassuring words of former president Dwight Eisenhower, who once said: "There is nothing wrong with America that the faith, love of freedom, intelligence and energy of her citizens cannot cure."[3]

Afterword

by Col. Douglas Macgregor

AMERICANS WHO LOOK CLOSELY at Roger Thompson's exceptional treatise on the U.S. Navy will discover an institution that persists in defending a set of modernization programs and operational concepts that rest on long-established habits of mind, social organization, and conventions, conditioned by the U.S. Navy's historical experience. The resulting indisposition over the last two decades inside the Navy's bureaucracy to recognize the significant threat posed to the Navy's institutional status quo by the combination of new technologies and new adversaries is not surprising. As always, conventional thinkers will try to dismiss Thompson's critique as irrelevant today, if only because the U.S. Navy has been lucky enough to avoid a wartime challenge since the end of World War II. Thankfully, though, not everyone agrees with this conventional "Nothing bad happened, and our main enemy is gone, so we must have been doing things right" viewpoint.

Change in the Navy's condition has been glacial, but there is reason for optimism. The new Chief of Naval Operations, Adm. Michael G. Mullen, wants to increase the navy's fleet of combatants to 313 ships by 2020, reversing years of decline in naval forces designed to operate in today's strategically complex environment. Admiral Mullen's plan envisions a major shipbuilding program that would increase the 281-ship fleet by 32 vessels and cost more than $13 billion a year, $3 billion more than the current shipbuilding budget. How Admiral Mullen will achieve his objective is not yet clear, but he is determined not simply to build new versions of platforms that already exist. Mullen, an experienced surface combatant officer, is painfully aware of how the Navy's obsession with supercarriers and nuclear submarines has retarded the combat capability of the surface navy and forced the service to rely on allies for essential services. Unlike his predecessor, Mullen is reluctant to invest the American taxpayer's money in naval platforms that substitute information derived from the promise of network-centric warfare for firepower and greater survivability. At the same time, Mullen is sensitive to the fiscal restraints that a new era of economic stringency will impose.

One reason for Mullen's desire to move in new directions is the extraordinary complexity of today's strategic environment. On one hand, he is compelled to maintain a naval force that can still employ a mix of surface combatants and submarines to preserve and enhance U.S. sea control while, on the other hand, building and employing new platforms for naval missions in new regions against new kinds of adversaries. For instance, the U.S. and Japanese navies are cooperating in the development of a theater/tactical ballistic missile-defense capability for AEGIS-equipped ships and are making substantial progress. If successful, this cooperation will provide critical capabilities to counter the People's Republic of China's reported development of terminally guided warhead medium-range ballistic missiles (MRBMs) for attacking U.S. and Japanese naval battle groups.

In other areas, such as the Mediterranean, West Africa, and the Southeast Asian archipelago, Mullen knows that the U.S. Navy must offer the nation a new and different set of naval capabilities. U.S. European Command is on its way to becoming a predominantly maritime theater as U.S. ground forces on the European continent are reduced and, eventually, withdrawn. In this new environment, American national security interests in and around the European periphery will depend heavily on a U.S. naval presence. From the Malaga Straits to Australia, U.S. Navy platforms will combat piracy and Islamist terrorism. As the American government moves to make security and governance in oil-rich West Africa an explicit priority and develops a comprehensive policy approach to the region, the U.S. Navy will have to develop and build platforms and capabilities commensurate with the U.S. security needs. All of the Navy's new platforms will have to focus on providing an offshore naval presence that is nonintrusive but still potent enough to intervene flexibly on behalf of U.S. interests while also encouraging positive regional political and economic development.

AFTERWORD

However, platforms are not the only concern for the Navy. Mullen knows the Marine Corps is overstretched and simply cannot provide the manpower for the Fleet Marine Force and continue to meet its obligations on land in the war on terror. In this setting, Mullen is pondering the fielding of naval security forces constituted from the Navy's own ranks, forces organized and equipped to secure naval vessels and infrastructure and to work with the Navy's Special Operations Forces in missions near or on land. Supplanting the Marine Corps capabilities in areas like West Africa will not be easy, but the Navy is clearly on a course to do so, on a limited basis.

How Mullen will proceed is hard to say. There are some models. The emphasis that Adm. William S. Sims, first president of the Naval War College, placed on war games would be helpful today if the war games were conducted in a freewheeling and honest fashion. Gaming, however, is only one element of intellectual honesty that must permeate the Navy's senior ranks if the Navy is to chart a new, more realistic path through the difficult and tumultuous waters that lie ahead. Mullen will also have to reward innovation in the shipbuilding industry by adopting new forms of propulsion and new ship designs.

Overcoming the bureaucratic reluctance to change the Navy with changing conditions will make necessary a return to the practice of eliminating unneeded bureaucracy and senior officers who man the bureaucracy. Thompson's example of the ruthless personnel changes implemented inside the U.S. Navy's World War II submarine force spring to mind, but even these measures may not be enough for the Navy. A reform and reorganization on the scale of the March 1942 executive order authorizing Gen. George C. Marshall to change the War Department fundamentally may well be necessary. As in all things in democratic society, whether Admiral Mullen receives the backing that he requires will be a function of support and backing from civilian appointees.

In his work *On Watch*, Adm. Elmo Zumwalt described the conditions that were essential to change, conditions that he was unable to achieve during his tenure as Chief of Naval Operations. These conditions included support from the Navy's sailors and junior officers for a vision of the Navy's future that was both tangible and attainable. In addition, the promotion of new senior officers with the understanding and desire to implement the new vision and the understanding and intelligent backing of the executive and legislative branches were also vital. It remains to be seen whether Admiral Mullen will achieve these conditions, but Thompson's work certainly provides useful inspiration for the process.

Douglas A. Macgregor, Ph.D.
Colonel, USA (Ret.)
Author of *Breaking the Phalanx*,
Transformation under Fire,
and *America's Lost Victory*

Washington, D.C.
January 4, 2006

Appendix: USN Ships That Have Been Theoretically Destroyed

Note: This table draws on information concerning unscripted exercise evolutions or operations as reported in the media since 1959. It is based on publicly available English media sources, such as newspapers, magazines, books, journals, and broadcast media. Only ships that have been named have been included.

Ship	Type	Date	Attacker
USS *Enterprise*	Aircraft carrier	???	Australian *Oberon*-class submarine
USS *Saratoga*	Aircraft carrier	1966	Soviet nuclear submarine *K-181* (9 simulated torpedo attacks)
USS *Independence*, *JFK*, and *FDR* (probable kills)	Aircraft carriers	1973	Soviet Navy's 5th Eskadra
USS *Eisenhower* and USS *Forrestal*	Aircraft carriers	1981	Canadian submarine and NATO submarine
USS *John F. Kennedy*	Aircraft carrier	1983	Canadian submarine
USS *Independence*	Aircraft carrier	1996	Chilean submarine
USS *Constellation*	Aircraft carrier	1997	Russian nuclear submarine

APPENDIX

Ship	Type	Date	Attacker
USS *Carl Vinson* (probable kill)	Aircraft carrier	1997	Russian nuclear submarine (apparently undetected)
USS *Theodore Roosevelt*	Aircraft carrier	1999	Dutch submarine
USS *Kitty Hawk*	Aircraft carrier	2000	Russian air force
USS *Abraham Lincoln*	Aircraft carrier	2000	Australian submarine
USS *Kitty Hawk* (probable kill)	Aircraft carrier	2006	Chinese submarine
USS *Skipjack* (probable kill)	Nuclear submarine	1959	Canadian Tracker aircraft
USS *Augusta* (probable kill)	Nuclear submarine	1986	Soviet nuclear submarine *K-219*
USS *Plunger*	Nuclear submarine	1988	Japanese submarine
USS *Hartford*	Nuclear submarine	1996	Lost 6 of 7 engagements against Canadian submarine
USS *Boise*	Nuclear submarine	1999	Dutch submarine
USS *Montpelier*	Nuclear submarine	2001	Chilean submarine
USS *Olympia*	Nuclear submarine	2002	Australian submarine
USS *City of Corpus Christi*	Nuclear submarine	2003	Australian submarine
USS *Pasadena*	Nuclear submarine	2003	Indian frigate
USS *Charlotte* (probable kill)	Nuclear submarine	2004	Canadian CP-140 aircraft
USS *Iowa*	Battleship	1988	Dutch frigate
USS *Iowa*	Battleship	1989	British, Canadian, and West German units
USS *Mount Whitney*	Command ship	1999	Dutch submarine

Notes

CHAPTER 1
1. Epigraph from Robert I. Fitzhenry, ed., *The Harper Book of Quotations* (New York: Quill, 1993), 452.
2. Michael Parenti, *Super Patriotism* (San Francisco: City Lights Books, 2004), 34.
3. Mickey Z., *The Seven Deadly Spins* (Monroe, Me.: Common Courage, 2004), 119.
4. Fitzhenry, *Harper Book of Quotations*, 78.
5. Skip Bowman, Speech "Honoring Our Own Greatest Generation" Chattanooga Armed Forces Day Luncheon. Chattanooga, Tennessee, May 6, 2005.
6. W. J. Holland, ed., *The Navy* (Washington, D.C.: Hugh Lauter Levin, 2000), 203.
7. Michael T. Isenberg, *Shield of the Republic: The United States Navy in an Era of Cold War and Violent Peace* (New York: St. Martin's, 1993), 1: 571, 829.
8. Robert L. O'Connell, *Sacred Vessels: The Cult of the Battleship and the Rise of the U.S. Navy* (New York: Oxford University Press, 1993), 80.
9. David Adams, "We Are Not Invincible," Naval Institute *Proceedings* (May 1997): 35–39.
10. *BrainyQuote*, www.brainyquote.com/quotes/authors/m/margaret_atwood.html.
11. Zenji Orita and Joseph Harrington, *I-Boat Captain* (Canoga Park, California: Major Books, 1976), 123.

CHAPTER 2
1. *Department of Defense Dictionary of Military and Associated Terms*, 120, available at www.dtic.mil/doctrine/jel/new_pubs/jp1_02.pdf.
2. Bradley Peniston, *Around the World with the U.S. Navy: A Reporter's Travels* (Annapolis, Md.: Naval Institute Press, 1999), 30.
3. *Department of Defense Dictionary of Military and Associated Terms*, 218.
4. Robert Coram, *Boyd: The Fighter Pilot Who Changed the Art of War* (New York: Back Bay Books, 2004), 384.

5. "Tandem Thrust '99," *Asia-Pacific Defense Forum* (online) (Fall 1999).
6. Mike Morley, "Kitty Hawk Packs a Punch in Tandem Thrust '99," COMVAVAIRPAC Public Affairs Office Report, March 30, 1999.
7. Tom Clancy, *Carrier* (New York: Berkley Books, 1999), 263.
8. Bill Gertz, "China Sub stalked U.S. fleet," *The Washington Times* (online), November 13, 2006.
9. Osamu Tagaya, *Imperial Japanese Naval Aviator 1937–45* (Wellingborough, U.K.: Osprey, 2003), back cover.
10. Thomas X. Hammes, *The Sling and the Stone* (St. Paul, Minn.: Zenith, 2004), ix.
11. *War Losses*, www.nwc.navy.mil/usnhdb/losses_war.asp.
12. Zenji Orita and Joseph Harrington, *I-Boat Captain* (Canoga Park, Calif.: Major Books, 1976), 123.
13. Ronald Spector, *At War at Sea* (New York: Penguin Books, 2002), 140.
14. James P. Levy, "Race for the Decisive Weapon: British, American, and Japanese Carrier Fleets 1942–1943," *Naval War College Review* 58, no. 1 (Winter 2005).
15. "A Brief History of Aircraft carriers: USS *Saratoga* (CV-3)," www.chinfo.navy.mil/navpalib/ships/carriers/histories/cv03-saratoga/cv03-saratoga.html.
16. Paul Ciotti, "Clinton's War on the Navy," *WorldNetDaily*, February 14, 2000.
17. Stan Goff, *Full Spectrum Disorder: The Military in the New American Century* (Brooklyn, N.Y.: Soft Skull, 2004), 116.
18. David Hackworth, *Hazardous Duty* (New York: Avon Books, 1996), 295.
19. Gordon Prange, *At Dawn We Slept* (New York: Penguin, 1991), 389.
20. Bruce M. Russett, *No Clear and Present Danger* (Boulder, Colo.: Westview, 1997), 22.
21. James Fallows, *National Defense* (New York: Random House, 1981), xv.
22. Richard Compton-Hall, *Submarine versus Submarine* (New York: Orion Books, 1988), 23.
23. Robert Fitzhenry, ed., *The Harper Book of Quotations* (New York: Quill 1993), 449.

CHAPTER 3

1. Richard Compton-Hall, *Submarine versus Submarine* (New York: Orion Books, 1988), 23.
2. John Benedict, "The Unraveling and Revitalization of U.S. Navy Antisubmarine Warfare," *Naval War College Review* 58, no. 2 (Spring 2005): 100.
3. Carl Hoffman, "Rotary Club" *Air & Space/Smithsonian* (January 2006): 22.
4. "Operation Mainbrace," *Time,* September 22, 1952.
5. Ibid.

6. Peter Hayes, Lyuba Zarsky, and Walden Bello, *American Lake: Nuclear Peril in the Pacific* (New York: Penguin Books, 1986), 69.
7. Mike Moore, "The Able–Baker–Where's Charlie Follies," *Bulletin of the Atomic Scientists,* May 1, 1994.
8. James P. Delgado, "Diving at Ground Zero: The Ships of Bikini Atoll Are a Grim Reminder of the Atomic Age," www.jamesdelgado.com/Diving_at_Ground_Zero.pdf.
9. Thomas B. Allen, *War Games* (New York: McGraw-Hill, 1987), 287.
10. Dean Knuth, "The Lessons of Ocean Venture '81," unpublished manuscript, 1981.
11. Morton Mintz, "Article Critical of Carriers Stamped 'Secret' by Navy," *Washington Post,* May 4, 1982, 1.
12. Michael T. Isenberg, *Shield of the Republic: The United States Navy in an Era of Cold War and Violent Peace* (New York: St. Martin's, 1993), 1: 584.
13. Dean Knuth, telephone interview with author, March 10, 2005.
14. Ibid.
15. Ibid.
16. Isenberg, *Shield of the Republic,* 441.
17. Ibid., 284, 721.
18. Knuth, "The Lessons of Ocean Venture '81."
19. Ibid.
20. Ibid.
21. James Fallows, *National Defense* (New York: Random House, 1981), 12.
22. Knuth, "The Lessons of Ocean Venture '81."
23. Ibid.
24. Ibid.
25. Ibid.
26. Ibid.
27. Hayes, Zarsky, and Bello, *American Lake,* 165.
28. "Carrier Group Five," www.globalsecurity.org/military/agency/navy/cargru5.htm.
29. "U.S. Flotilla Decked by Canadian Sub," *Winnipeg Free Press,* May 9, 1984, 30.
30. Compton-Hall, *Submarine versus Submarine,* 184.
31. Knuth, "The Lessons of Ocean Venture '81."
32. "Zwaardvis," www.dutchsubmarines.com/boats/boat_zwaardvis2.htm.
33. "Walrus," www.dutchsubmarines.com/boats/boat_walrus2.htm.
34. "Vice Admiral C. A. Lockwood, ComSubPac," Dutchsubmarines.com, www.dutchsubmarines.com/specials/special_lockwood.htm.
35. "Aussie Duo Conquer Perisher Challenges," Navy News (November 2002), repr. Dutchsubmarines.com, www.dutchsubmarines.com/specials/special_aussie_duo.htm.
36. Ibid.
37. Todd Cloutier, "Daring to Go Dutch," *Undersea Warfare Magazine* (online) (Fall 2003).

38. Ibid.
39. Corwin Mendenhall, "The Case for Diesel Submarines," Naval Institute *Proceedings*, letter (September 1995), 27.
40. Ivan Eland, "It's Time to Mothball Unneeded Nuclear Attack Submarines," *USA Today* (magazine), September 1, 1999.
41. Jennifer Taplin, "Flood of Memories about Go Boat Sub," *Halifax Daily News* (online), March 20, 2006.
42. Peter Kavanagh, e-mail to author, March 23, 2006.
43. "Oberon Class Submarine," en.wikipedia.org/wiki/Oberon_class_submarine.
44. Brendan Nicholson, "*Collins* Subs Star in Naval Exercises," *The Age* (online), September 24, 2003.
45. Nathan Hodge, "Australian Hit on U.S. Sub Gets Attention," *Defense Week Daily Update*, October 1, 2003.
46. "On the Defensive," *Bulletin of the Atomic Scientists*, January 1, 2004.
47. Compton-Hall, *Submarine versus Submarine*, 137.
48. Brendan Nicholson, "*Collins* Sub Shines in U.S. War Game," *The Age*, October 13, 2002.
49. Derek Woolner, "Getting in Early: Lessons of the *Collins* Submarine Program for Improved Oversight of Defence Procurement," Research Paper 3 2002-02 (Canberra, ACT: Department of the Parliamentary Library, Foreign Affairs, Defence and Trade Group, 18 September 2000, 22.
50. Maryanne Kelton, "New Depths in Australia-U.S. Relations: The *Collins* Class Submarine Project," School of Political and International Studies Working Paper, Flinders University of South Australia, March 2004, 22–23.
51. Brendan Nicholson, "Sub Troops Get Underwater Cover," *The Age*, March 2, 2003.
52. E-mail from Capt. Jan Nordenman, June 28, 2005.
53. E-mail from Andy Karam, May 2, 2005.
54. Ronald Spector, *At War at Sea* (New York: Penguin, 2001), 155.
55. "Defense Watch," *Defense Daily*, June 18, 2001.
56. Chris Lambie, "Fast Torpedoes Pose New Threat," *Daily News* (online), August 6, 2004.
57. Jean Morin and Richard Gimblett, *Operation Friction* (Toronto: Dundurn, 1997), 85–88.
58. Richard J. Newman, "Breaking the Surface," *U.S. News & World Report*, April 6, 1998.
59. Robert Holzer, "Dangerous Waters: Submarines, New Mines Imperil Ill-Prepared U.S. Navy Fleet," *Defense News*, May 4–10, 1998, 1, 14–15. Quoted in Shirley A. Kan et al., "China's Foreign Conventional Arms Acquisitions: Background and Analysis," *Congressional Research Service Report*, October 10, 2000, 64.

60. Benedict, "The Unraveling and Revitalization of U.S. Navy Antisubmarine Warfare," 60.
61. Newman, "Breaking the Surface."
62. "Defense Watch."
63. Andrew Cockburn, *The Threat: Inside the Soviet Military Machine* (New York: Vintage Books, 1984), 426.
64. Ashley Bennington, "Stealthy Subs," letter, *Popular Science,* July1, 1999.
65. E-mail from Andy Karam, 13 April 2005.
66. Ibid.
67. Michael DiMercurio and Michael Benson, *The Complete Idiot's Guide to Submarines* (New York: Alpha Books, 2003), 162.
68. Ibid., 56.
69. Compton-Hall, *Submarine versus Submarine,* 30.
70. Viktor Toyka, "Ask Questions about Our Ability to Conduct Anti Submarine Warfare," Naval Institute *Proceedings* (September 2004), 24.
71. Brian Hsu, "Navy Allows a Rare Glimpse of Sub," *Taipei Times,* June 23, 2000.
72. Robert Williscroft, "Tomorrow's Submarine Fleet: The Non-nuclear Option," *DefenseWatch,* February 2, 2002.
73. Tom Clancy, *Submarine* (New York: Berkley Books, 1993), 149.
74. E-mail from John Byron. September 20, 2005.
75. Scott Shuger, "The Navy We Need and the One We Got," *Washington Monthly* (March 1989).
76. Robert Fitzhenry, ed., *The Harper's Book of Quotations* (New York: Quill, 1993), 140.
77. E-mail from Andy Karam, 13 April 2005.
78. Shuger, "The Navy We Need and the One We Got."
79. Ed Marolda, "Mine Warfare," www.history.navy.mil/ wars/korea/ minewar.htm.
80. David Hackworth, *Hazardous Duty* (New York: Avon, 1996), 201.
81. Duncan Miller and Sharon Hobson, *The Persian Excursion* (Clementsport, N.S.: Canadian Peacekeeping, 1995), 87.
82. Ibid.
83. Edward Marolda and Robert Schneller, Jr., *Shield and Sword* (Washington, D.C.: Government Reprints, 2001), 77.
84. Ibid., 254.
85. Ibid., 263.
86. Ibid., 149.
87. Ibid., 322.
88. Frank G. Coyle, "Navy Needs Heavy-Lift/Countermine Helos," Naval Institute *Proceedings* (August 2004): 52–54.
89. Paul Ryan, "LCS Will Transform Mine Warfare," Naval Institute *Proceedings* (December 2004): 38.
90. Ibid.

CHAPTER 4

1. John Benedict, "The Unraveling and Revitalization of U.S. Navy Antisubmarine Warfare," *Naval War College Review* 58, no. 2 (Spring 2005): 99–100.
2. George C. Wilson, *Supercarrier* (New York: Berkley Books, 1988), 236.
3. E-mail from Leif Wadelius, April 13, 2005.
4. E-mail from Andy Karam, May 24, 2005.
5. G. Voronin, "The Silence of Our Submarines Annoys Not Only Dilettantes," *Krasnaya zvezda,* 28 January 1994, 5.
6. Owen Cote, *The Third Battle: Innovation in the U.S. Navy's Cold War Struggle with Soviet Submarines* (Cambridge, Mass.: MIT Security Studies Program, March 2000).
7. Ibid.
8. Robert Moore, *A Time to Die* (New York: Three Rivers, 2002), 117–18.
9. E-mail from Andy Karam, December 7, 2005.
10. Norman Polmar, "Statement by Norman Polmar" to the House Military Procurement Subcommittee, Washington, D.C., March 18, 1997.
11. Michael DiMercurio and Michael Benson, *The Complete Idiot's Guide to Submarines* (New York: Alpha Books, 2003), 23.
12. Christy Tillery French, interview with Michael DiMercurio, available at www.ussdevilfish.com/interv.htm.
13. Polmar, "Statement."
14. Tom Clancy, *Carrier* (New York: Berkley Books, 1999), 45.
15. Alan E. Diehl, *Silent Knights* (Dulles, Va.: Brassey's, 2002), 186.
16. Lyle Goldstein and William Murray, "Undersea Dragons: China's Maturing Submarine Force," *International Security* 28, no.4 (Spring 2004): 184.
17. Dave Mayfield, "Subs Emerge from the Deep," *Virginian-Pilot,* April 27, 1998.
18. Robert J. Caldwell, "Navy's Woes Reflect Risks of Years of Underfunding Defense," *San Diego Union Tribune* (online), September 24, 2000.
19. James W. Crawley, "S.D. Might Be Home to Swedish Sub," *San Diego Union-Tribune,* October 14, 2004.
20. "Svensk ubat gackar marinen I U.S.A.," *Gefle Dagblad* (online), 5 October 2005.
21. Norman Polmar and K. J. Moore, *Cold War Submarines* (Dulles, Va.: Brassey's, 2004), 210–14.
22. Nickolay Nedelchev, "The US Silent Service in Early WW II: A Story of Failure," uboat.net/allies/technical/torpedo_problems.htm.
23. Michael Gannon, *Operation Drumbeat: Germany's U-Boat Attacks along the American Coast in World War II* (New York: HarperPerennial, 1991), 309.
24. Ibid., 338.

25. Peter Padfield, *War beneath the Sea* (New York: John Wiley and Sons, 1998), 194.
26. Homer Hickam, *Torpedo Junction* (New York: Dell, 1989), 173.
27. Eliot A. Cohen and John Gooch, *Military Misfortunes: The Anatomy of Failure in War* (New York: Anchor, 2003), 75.
28. Karl Doenitz, *Memoirs: Ten Years and Twenty Days* (New York: Da Capo, 1997), 202.
29. Andrew Williams, *The Battle of the Atlantic* (New York: Basic Books, 2003), 179.
30. Zenji Orita and Joseph Harrington, *I-Boat Captain* (Canoga Park, Calif.: Major Books, 1976), 40.
31. Ibid., 42.
32. Williams, *Battle of the Atlantic,* 169.
33. Richard Deacon, *The Silent War: A History of Western Naval Intelligence* (London: Grafton Books, 1988), 189–90.
34. Ibid., 190.
35. Jeffry Brock, *The Dark Broad Seas,* vol. 1, *With Many Voices* (Toronto: McClelland and Stewart, 1981), 92.
36. Cohen and Gooch, *Military Misfortunes,* 65.
37. John Reeve and David Stevens, eds., *The Face of Naval Battle* (Crows Nest, NSW: Allen and Unwin, 2003), 99–101.
38. Ibid.
39. Hickam, *Torpedo Junction,* 56.
40. Cohen and Gooch, *Military Misfortunes,* 66–67.
41. Dan van der Vat, *The Pacific Campaign* (New York: Simon and Schuster, 1991), 120.
42. Ibid., 270.
43. Gannon, *Operation Drumbeat,* 309.
44. Roger Sarty, "The Royal Canadian Navy and the Battle of the Atlantic, 1939–1945" (Ottawa: Canadian War Museum, 2001), www.civilization.ca/cwm/disp/dis007_e.html.
45. Gannon, *Operation Drumbeat,* 179.
46. Peter J. Dombrowski and Andrew L. Ross, "Transforming the Navy: Punching a Feather Bed?" *Naval War College Review* 56, no. 3 (Summer 2003).
47. Hickam, *Torpedo Junction,* 94.
48. Ibid.
49. Doenitz, *Memoirs,* 216.
50. Gannon, *Operation Drumbeat,* 389.
51. Samuel Eliot Morison. *The Battle of the Atlantic* (Urbana: University of Illinois Press, 2001), 303.
52. Gannon, *Operation Drumbeat,* 390–91.
53. Doenitz, *Memoirs,* 203.
54. Hickam, *Torpedo Junction,* xi.

55. Ibid., 64.
56. Ibid., 31.
57. Gannon, *Operation Drumbeat*, 378–79.
58. Frank Hoffman, "Thinking Ahead Intelligently," review of *Uncovering Ways of War: U.S. Intelligence and Foreign Military Innovation*, by Thomas C. Mahnken, *Naval War College Review* 56, no. 1 (Winter 2003): 157.
59. Homer Hickam, *Torpedo Junction* (New York: Dell, 1989), xii.
60. Nathan Miller, *War at Sea* (New York: Oxford University Press, 1995), 193.
61. Morison, *Battle of the Atlantic*, 214.
62. Douglas M. Brattebo, "From the Guest Editor," *White House Studies* (Spring 2004).
63. Morison, *Battle of the Atlantic*, 219.
64. Michael DiMercurio and Michael Benson, *The Complete Idiot's Guide to Submarines* (New York: Alpha, 2003), 299.
65. Brock, *The Dark Broad Seas*, 83.
66. Ibid.
67. Williams, *Battle of the Atlantic*, 176.
68. James J. Sadkovich, ed., *Reevaluating Major Naval Combatants of World War II* (Westport, Conn.: Greenwood, 1990), xx, 10.
69. Brock, *The Dark Broad Seas*, 135.
70. W. G. D. Lund. "The Royal Canadian Navy's Quest for Autonomy in the North West Atlantic: 1941–1943," in *RCN in Retrospect 1910–1968*, ed. James A. Boutilier (Vancouver: University of British Columbia Press, 1982), 156.
71. Ibid., 155.
72. Ibid., 146.
73. Ibid.
74. Ibid., 155.
75. Geoffrey Regan, *The Brassey's Book of Naval Blunders* (Washington, D.C.: Brassey's, 2000), 116–17.
76. Ibid.
77. Sadkovich, ed., *Reevaluating Major Naval Combatants of World War II*, xx, 10.
78. Morison, *Battle of the Atlantic*, 12–13.
79. Imperial War Museum, www.iwm.org.uk/upload/package/8/atlantic/can3942.htm.
80. Spector, *At War at Sea*, 252.
81. Dan van der Vat, *The Atlantic Campaign* (Edinburgh: Birlinn, 2001), 13.
82. Ibid., 444.
83. Ibid., 381.
84. Ibid., 383.
85. Ibid., 447.
86. Tony German, *The Sea Is at Our Gates* (Toronto: McClelland and Stewart, 1991), 111.

87. Marc Milner, *North Atlantic Run: The Royal Canadian Navy and the Battle for the Convoys* (Toronto: University of Toronto Press, 1994).
88. Brock, *The Dark Broad Seas,* 91.
89. Ibid.
90. Ibid., 98.
91. Ibid.
92. Ibid, 197–98.
93. Peter Huchthausen, *October Fury* (Hoboken, N.J.: John Wiley and Sons, 2002), 113.
94. Ibid.
95. Michael T. Isenberg, *Shield of the Republic: The United States Navy in an Era of Cold War and Violent Peace* (New York: St. Martin's, 1993), 364.
96. German, *The Sea Is at Our Gates,* 236.
97. Williams, *Battle of the Atlantic,* 173.
98. Benedict, "The Unraveling and Revitalization of U.S. Navy Antisubmarine Warfare," 98.
99. Stuart E. Soward, *Hands to Flying Stations* (Victoria, B.C.: Neptune Developments, 1995), 2: 187–91.
100. Isenberg, *Shield of the Republic,* 691.
101. David Robinson, "The Canadian Angle during the Cuban Missile Crisis," www.alts.net/ns1625/nshist21.html.
102. John Ward, "Admiral Unilaterally Sent Canadian Warships as Part of U.S. Maritime Blockade of Cuba," *Shunpiking,* www.shunpiking.com/nhfw/cuba-can-blockade.htm.
103. German, *The Sea Is at Our Gates,* 266–73.
104. Ibid.
105. Peter Haydon, *The 1962 Cuban Missile Crisis: Canadian Involvement Reconsidered* (Toronto: Canadian Institute of Strategic Studies, 1993), 26.
106. Brock, *The Thunder and the Sunshine,* 112.
107. Huchthausen, *October Fury,* 91–92.
108. Ibid.
109. Ibid.
110. Robert Gordon, "Navy's Ships Making Valuable Contribution," *Halifax Chronicle Herald,* March 23, 1983, 3.
111. E-mail from Andy Karam, 13 April 2005.
112. Michael Abrashoff, *It's Your Ship* (New York: Warner Books, 2002), 16–17.
113. "CDC's CANTASS Is an Exceptional Canadian Achievement," *Wednesday Report* (online) 4, no. 36.
114. German, *The Sea Is at Our Gates.*
115. Richard Compton-Hall. *Submarine versus Submarine* (Toronto: Collins, 1988), 35.
116. David Miller, *The Illustrated Directory of Submarines of the World* (London: Salamander, 2002), 272.

117. Jonathan Powis, "U.K.'s *Upholder*-Class Subs Go to Canada," Naval Institute *Proceedings* (October 2002): 101.
118. Gerry Pash, "Rim of the Pacific 2004: Aurora Hunts U.S.N. Submarine," *Lookout,* July 26, 2004.
119. Michael DiMercurio and Michael Benson, *The Complete Idiot's Guide to Submarines* (New York: Alpha, 2003), 47.
120. "Buy Six Diesel-Electric Submarines for Antisubmarine Warfare Training (050-26)" in *Budget Options 2001* (online) (Washington, DC: Congressional Budget Office, February 2001).

CHAPTER 5

1. *BrainyQuote,* www.brainyquote.com/quotes/quotes/l/leftygomez139632.html.
2. Paul Ciotti, "Clinton's War on the Navy," *WorldNetDaily,* February 14, 2000.
3. Interview with Thomas B. Allen, *Talk of the Nation* with Neal Conan, NPR, January 8, 2003.
4. Stephanie Gutmann, *The Kinder, Gentler Military* (San Francisco: Encounter Books, 2000), 23.
5. Ibid., 90–91.
6. James F. Dunnigan and Albert A. Nofi, *Dirty Little Secrets of the Vietnam War* (New York: Thomas Dunne, 1999), 145.
7. Image of North Vietnamese postage stamp commemorating the sinking of USS *Card,* www.navsource.org/archives/03/0301132.jpg.
8. Dunnigan and Nofi.
9. Christopher Cerf and Victor Navasky, *The Experts Speak* (New York: Villard, 1998), 266.
10. Dan van der Vat, *The Pacific Campaign* (New York: Touchstone, 1991), 351.
11. Robert I. Fitzhenry, ed., *The Harper Book of Quotations* (New York: Quill, 1993), 140.
12. Scott Shuger, "The Navy We Need and the One We Got," *Washington Monthly* (March 1989).
13. Joseph Enright and James Ryan, *Sea Assault* (New York: St. Martin's, 2000), 17, 28.
14. Thinkexist, en.thinkexist.com/quotation/power_is_not_revealed_by_striking_hard_or_often/11063.html.
15. Robert Williscroft, "Lessons from Two Aging Aircraft Carriers," *DefenseWatch,* October 9, 2002.
16. Ibid.
17. Ibid.
18. Rick Chernitzer. "Problems Corrected, USS John F Kennedy ready to go," *Star and Stripes* (online), February 25, 2002.
19. Geoffrey Regan, *The Brassey's Book of Naval Blunders* (Dulles, Va.: Brassey's, 2000), 49.

20. Ibid.
21. Charles C. Thompson II, *A Glimpse of Hell: The Explosion on the USS* Iowa *and Its Cover-up* (New York: W. W. Norton, 1999), 181.
22. Ibid., 68.
23. Ibid., 74.
24. Peter Cary, "Death at Sea," *U.S. News & World Report,* April 23, 1990.
25. Regan, *Brassey's Book of Naval Blunders,* 84–85.
26. Theodore F. Cook. "Our Midway Disaster: Japan Springs a Trap, June 4, 1942," in *What If? The World's Foremost Military Historians Imagine What Might Have Been,* ed. Robert Cowley (New York: Berkley Books, 1999), 311–12.
27. Ibid.
28. Samuel E. Morison, *The Rising Sun in the Pacific 1931–April 1942* (Edison, N.J.: Castle Books, 2001), 6, 23–25.
29. Ibid.
30. Ibid.
31. Robert L. O'Connell, *Sacred Vessels: The Cult of the Battleship and the Rise of the U.S. Navy* (New York: Oxford University Press, 1991), 311–12.
32. Zenji Orita and Joseph D. Harrington, *I-Boat Captain* (Canoga Park, Calif.: Major Books, 1976), 73.
33. Mitsuo Fuchida and Masatake Okumiya, *Midway: The Battle That Doomed Japan, the Japanese Navy's Story* (Annapolis, Md.: Naval Institute Press, 2001) , 273, 275–76,279.
34. O'Connell, *Sacred Vessels,* 312.
35. Walter Lord, *Incredible Victory: The Battle of Midway* (Short Hills, N.J.: Burford Books, 1998), ix.
36. Ibid., 147.
37. Alvin Kernan, *The Unknown Battle of Midway: The Destruction of the American Torpedo Squadrons, June 4, 1942* (New Haven, Conn.: Yale University Press, 2005), 41, 45–46, 112.
38. Ibid., 25.
39. Ibid., xiv, 39.
40. Ibid., 46.
41. Ibid., 128–31.
42. Tom Clancy, *Carrier* (New York: Berkley Books, 1999), 56.
43. S. G. Gorshkov, *The Sea Power of the State* (Oxford: Pergamon, 1980), 114.
44. Orita and Harrington, *I-Boat Captain,* 123–24.
45. Ibid., 125.
46. Richard Overy, *Why the Allies Won* (New York: W. W. Norton, 1995), 38.
47. Forrest R. Lindsey, "Nagumo's Luck," in *Rising Sun Victorious,* ed. Peter G. Tsouras (London: Greenhill Books: 2001), 124.
48. Peter Padfield, *War beneath the Sea* (New York: John Wiley and Sons, 1998), 191.
49. Ronald H. Spector, *Eagle against the Sun* (New York: Vintage, 1985), 20.
50. Brayton Harris, *The Navy Times Book of Submarines* (New York: Berkley Books, 1997), 311–12.

51. Richard Compton-Hall, *Submarine vs. Submarine* (Toronto: Collins, 1988), 107.
52. Orita and Harrington, *I-Boat Captain*, 200.
53. J. L. Granatstein, "The American Influence on the Canadian Military, 1939–1963" *Canadian Military History* (online) (Spring 1993).
54. Lindsey, "Nagumo's Luck," 125.
55. Orita and Harrington, *I-Boat Captain*, 201.
56. Vox Day, "No case for Internment," *WorldNetDaily,* September 6, 2004.
57. Roger Bruns, *Almost History* (New York: Hyperion, 2000), 233.
58. Scott Shuger, "Hurts So Good: At POW School the Navy Locked Me in a Box," *Washington Monthly* (May 1988).
59. Jonathan Neale, *A People's History of the Vietnam War* (New York: New Press, 2003), 75.
60. Christopher Cerf and Victor Navasky, *The Experts Speak* (New York: Villard, 1998), 135.
61. Thomas Hammes, *The Sling and the Stone* (St. Paul, Minn.: Zenith, 2004), viii.
62. Norman Dixon, *On the Psychology of Military Incompetence* (London: Pimlico, 1994), 349–50.
63. Robert Coram, *Boyd: The Fighter Pilot Who Changed the Art of War* (New York: Back Bay Books, 2002), 377.
64. Gordon Prange et al., *At Dawn We Slept* (New York: Penguin, 1991), 392.
65. Coram, *Boyd,* 331–32.
66. Rick Beyer, *The Greatest War Stories Never Told* (New York: Collins, 2005), 2–3.
67. Coram, *Boyd,* 377.
68. Beyer, *Greatest War Stories Never Told,* 30–31.
69. Ibid., 56–57.
70. Stephen Biddle, *Military Power: Explaining Victory and Defeat in Modern Battle* (Princeton, N.J.: Princeton University Press, 2004), ix.
71. Ibid., 23.
72. Lord, *Incredible Victory,* 3.
73. "Automobile Production, United States, Japan and Germany, 1950–2004 (in millions)," people.hofstra.edu/geotrans/eng/ch1en/conc1en/carprod1950-1999.html.
74. William Greider, *Fortress America* (New York: PublicAffairs, 1998), 138.
75. Victor N. Corpus, "If It Comes to a Shooting War," *Asia Times Online,* April 20, 2006.
76. Thinkexist, en.thinkexist.com/quotation/it-s_not_the_size_of_the_dog_in_the_fight-it-s/14187.html.
77. Overy, *Why the Allies Won,* 38, 298–301, 324.
78. Bruce M. Russett, *No Clear and Present Danger: A Skeptical View of the United States Entry into World War II* (Boulder, Colo.: Westview, 1997), 22, 25–27, 49–62.
79. Edwin Hoyt, *War in the Pacific,* vol. 1, *Triumph of Japan* (New York: Avon Books, 1990), 35–36.

80. Ibid.
81. Dan van der Vat, *The Pacific Campaign: The U.S.-Japanese Naval War 1941–1945* (New York: Touchstone, 1991), 19.
82. Elihu Rose, "The Case of the Missing Carriers," in *What If? The World's Foremost Military Historians Imagine What Might Have Been,* ed. Robert Cowley (New York: Berkley Books, 1999), 340.
83. Eric Margolis, "Did Russia Win D-Day?" *CNEWS,* May 30, 2004.
84. S. G. Gorshkov, *The Sea Power of the State* (Oxford: Pergamon, 1980), 117.

CHAPTER 6

1. Patrick Tyler, *Running Critical: The Silent War, Rickover, and General Dynamics* (New York: Harper and Row, 1986), 49
2. Jon Dougherty, "Russian Flyover Takes Navy by Surprise," *WorldNetDaily,* December 7, 2000.
3. "U.S. Pacific Fleet No Match for the Russian Air Force," *Discerning the Times Digest and Newsbytes* (November 2000).
4. Kick, Russ, ed., *Everything You Know Is Wrong* (New York: Disinformation, 2003), 221–22.
5. Edward Timperlake and William C. Triplett II, *Red Dragon Rising: Communist China's Military Threat to America* (Washington, D.C.: Regnery, 2000), 166–67.
6. Tom Clancy, *Carrier* (New York: Berkley Books, 1999), 263.
7. Lauren Holland, *Weapons under Fire* (New York: Garland, 1997), 139.
8. George C. Wilson, *Supercarrier* (New York: Berkley Books, 1986), 281–82.
9. James P. Stevenson, *The Pentagon Paradox: The Development of the F-18 Hornet* (Annapolis, Md.: Naval Institute Press, 1993), 114.
10. Gregory L. Vistica, *Fall from Glory: The Men Who Sank the U.S. Navy* (New York: Simon and Schuster, 1997), 157.
11. Dougherty, "Russian Flyover Takes Navy by Surprise."
12. Robyn Dixon and Paul Richter, "Russia Brags Its Jets Buzzed Carrier *Kitty Hawk,*" *Virginian-Pilot,* November 16, 2000.
13. *Defense Acquisitions: Comprehensive Strategy Needed to Improve Ship Cruise Missile Defense,* Report GAO/NSIAD-00-0149 (Washington, D.C.: GAO, July 2000).
14. Sean Paige, "GAO Offers Stark Assessment of Warship Vulnerability," *Insight on the News,* August 14, 2000.
15. Jeffery L. Levinson and Randy L. Edwards, *Missile Inbound: The Attack on the* Stark *in the Persian Gulf* (Annapolis, Md.: Naval Institute Press, 1997), 100.
16. Everest Riccioni, *Is the Air Force Spending Itself into Unilateral Disarmament?* (Washington, D.C.: Project on Government Oversight, August 2001).
17. Ibid.

18. Chester Richards, *A Swift, Elusive Sword* (Washington, D.C.: CDI, 2003), 47.
19. James Fallows, *National Defense* (New York, Random House, 1981), 41–42.
20. Tom Cooper and Farzad Bishop, *Iranian F-14 Tomcat Units in Combat* (Oxford: Osprey, 2004), 17, 32, 35, 44, 82, 92.
21. David Isenberg, *The Illusion of Power: Aircraft Carriers and U.S. Military Strategy*, Cato Policy Analysis 134 (Washington, D.C.: Cato Institute, 1990).
22. Scott Shuger, *Navy Yes, Navy No*, unpublished manuscript, 2.
23. Robert K. Wilcox, *Wings of Fury* (New York: Pocket Books, 2004), 13–14.
24. Ibid.
25. Ibid.
26. Lon O. Nordeen, "A Half Century of Jet-Fighter Combat: Skill, Determination, and Effective Battle Planning and Tactics Are at Least as Important as Excellent Birds of Prey," *Journal of Electronic Defense* (January 2004).
27. "Vympel R-77," en.wikipedia.org/wiki/Vympel_R-77.
28. Hugh McManners, *Top Guns* (London: Network Books, 1996), 167–77.
29. Rowan Scarborough, "U.S. Ship Took 40 Minutes to Respond to Order," *Washington Times*, December 7, 2000.
30. Shuger, *Navy Yes, Navy No*, 166.
31. E-mail from J. R Sampson, July 7, 2004.
32. Gary E. Weir and Walter J. Boyne, *Rising Tide: The Untold Story of the Russian Submarines That Fought the Cold War* (New York: Basic Books, 2003), 118.
33. Ibid., 117.
34. Patrick Tyler, *Running Critical: The Silent War, Rickover, and General Dynamics* (New York: Harper and Row, 1986), 25.
35. John Pina Craven, *The Silent War: The Cold War Battle beneath the Sea* (New York: Touchstone, 2001), 217.
36. Peter Hayes, Lyuba Zarsky, and Walden Bello, *American Lake: Nuclear Peril in the Pacific* (Middlesex: Penguin Books, 1987), 207.
37. Scott Shuger, "Paperback Fighter," *Washington Monthly* (November 1989).
38. Peter Truscott, *Kursk* (London: Pocket books, 2003), 124.
39. Sherry Sontag and Christopher Drew, *Blind Man's Bluff* (New York: HarperPaperbacks, 1998), 299.
40. Weir and Boyne, *Rising Tide*, 207–208.
41. Robert Kaylor. "The Navy's 21st-Century Submarine: Critics Ask Whether the *Seawolf* Can Keep the U.S. ahead of the Soviets," *U.S. News & World Report*, April 24, 1989.
42. Peter Huchthausen, Igor Kurdin, and R. Alan White, *Hostile Waters* (New York: St. Martin's Paperbacks, 1997), 42, 57, 59–68.
43. Truscott, *Kursk*, 130.

44. Bill Gertz, "Russian Sub Stalks Three U.S. Carriers," *Washington Times*, November 23, 1997.
45. Ibid.
46. Ibid.
47. Norman Polmar, statement to the House Military Procurement Subcommittee, Washington, D.C.: March 18, 1997

CHAPTER 7

1. Michael Pillsbury, *China Debates the Future Security Environment* (online) (Washington, D.C.: National Defense University Press, 2000).
2. Ibid.
3. Michael O'Hanlon, Lyle Goldstein, and William Murray, "Damn the Torpedoes: Debating Possible U.S. Navy Losses in a Taiwan Scenario," *International Security* 29, no. 2 (Fall 2004): 202–204.
4. Ibid., 205.
5. Ibid.
6. Victor N. Corpus, "If It Comes to a Shooting War," *Asia Times Online*, April 20, 2006.
7. Ibid.
8. Ted Galen Carpenter and Justin Logan, "Eastern Fronts," *American Prospect Online Edition*, May 16, 2005, www.prospect.org/web/page.ww?section=root&name=ViewWeb&articleId=9670.
9. Corpus, "If It Comes to a Shooting War."

CHAPTER 8

1. Toby Westerman, "Naval Officer Warns of Attack: Daly says Nuclear Sub, Carrier Bases Vulnerable," *WorldNetDaily*, June 19, 2001.
2. WABC-TV, New York. "Investigators Uncover Serious Flaws in Security at U.S. Navy Bases," October 24, 2000, transcript available at abclocal.go.com/wabc/features/WABC_investigators_navybases.html.
3. "About Dick Marcinko," www.dickmarcinko.com/AboutDM.aspx.
4. Richard Marcinko and John Weisman, *Rogue Warrior* (New York: Pocket Books, 1992), 338.
5. Ibid., 340.
6. Ibid., 341.
7. E-mail from Andy Karam, March 22, 2006.
8. Michael Hirsh, *None Braver: U.S. Air Force Pararescuemen in the War on Terrorism* (New York: NAL Caliber, 2004), 62.
9. Ibid., 80.
10. Hans Halberstadt, *U.S. Navy SEALS* (Osceola, Wis.: MBI, 1993), 83.
11. Donald E. Vandergriff, review of *Not a Good Day to Die: The Untold Story of Operation Anaconda*, www.d-n-i.net/vandergriff/naylor_review.htm.

CHAPTER 9

1. Andrew Cockburn, *The Threat: Inside the Soviet Military Machine* (New York: Vintage Books, 1984), 424–25.
2. Richard A. Stubbing and Richard A. Mendel, *The Defense Game* (New York: Harper and Row, 1986), 118.
3. Christopher J. Kelly, "The Submarine Force in Joint Operations." Research Report AU/ASCS/145-1998-04 (Maxwell Air Force Base, Ala.: Air Command and Staff College, April 1998), 20–21.
4. Perry M. Smith, *Assignment Pentagon* (Washington: Brassey's, 2002), 176.
5. Ronald Spector, *At War at Sea: Sailors and Naval Combat in the Twentieth Century* (New York: Penguin Books, 2001), 163.
6. Ibid.
7. Cockburn, *The Threat*, 424.
8. Gregory L. Vistica, *Fall from Glory: The Men Who Sank the U.S. Navy* (New York: Simon and Schuster, 1997), 91.
9. John Lehman, *Command of the Seas* (New York: Charles Scribner's Sons, 1988), 99.
10. Robert Kaylor, "Deadly Games of Hide and Seek," *U.S. News & World Report*, June 15, 1987.
11. Scott Shuger, "The Navy We Need and the One We Got," *Washington Monthly* (March 1989).
12. Dave Mayfield, "Subs Emerge from the Deep," *Virginian-Pilot*, April 27, 1998.
13. Stansfield Turner, "Is the U.S. Navy Being Marginalized?" *Naval War College Review* 56, no. 3 (Summer 2003).
14. Ibid.
15. Spector, *At War at Sea*, 376.
16. Edward J. Marolda and Robert J. Schneller, Jr., *Shield and Sword: The United States Navy and the Persian Gulf War* (Washington, D.C.: Government Reprints, 2001), 24–25.
17. Peter Hayes, Lyuba Zarsky, and Walden Bello, *American Lake: Nuclear Peril in the Pacific* (Middlesex: Penguin Books, 1987), 186.
18. Ibid.
19. Lyle Goldstein and Yuri M. Zhukov, "Superpower Showdown in the Mediterranean, 1973," *Sea Power* (online) (October 2003).
20. Lyle Goldstein and Yuri M. Zhukov, "A Tale of Two Fleets: A Russian Perspective on the 1973 Naval Standoff in the Mediterranean," *Naval War College Review* 57, no. 2 (Spring 2004).
21. Ibid.
22. "Wake-Up Call," *Guardian*, September 6, 2002, available at www.guardian.co.uk/g2/story/0,3604,786992,00.html.
23. Stan Goff, *Full Spectrum Disorder: The Military in the New American Century* (Brooklyn, N.Y.: Soft Skull, 2004), 73.
24. Stubbing and Mendel, *The Defense Game*, 122.

25. Douglas A. Macgregor, *Breaking the Phalanx* (Westport, Conn.: Praeger, 1997), 127.
26. Ibid., 142.
27. D. Michael Abrashoff, *Get Your Ship Together* (New York: Portfolio, 2004), 106–107.
28. Robert D. Williscroft, "The New Submarine Paradigm," thedeadhand.com/blogs/rgw/articles/640.aspx.
29. Macgregor, *Breaking the Phalanx*, 137.

CHAPTER 10
1. E-mail from Everest Riccioni, February 14, 2005.
2. Charles Moskos, "Armed Forces in a Warless Society," in *War*, ed. Lawrence Freedman (Oxford: Oxford University Press, 1994), 136–37.
3. "Canada's Air Force, History, World War I," www.airforce.gc.ca/hist/ww_1_e.asp,
4. "Facts That May Surprise You," www.canadianembassy.org/ca/facts-mil-en.asp.
5. Martin A. Noel, "Review of Knights of the Air: Canadian Fighter Pilots in the First World War," *Air & Space Power Journal* (Fall 2003).
6. John Keegan, *Six Armies in Normandy* (New York: Penguin, 1994), xviii, 121.
7. Ibid.
8. Rob Tate, "Malta Spitfire: The Diary of a Fighter Pilot" (book review), *Air & Space Power Journal* (online) (Spring 2004).
9. Dan McCaffrey, *Air Aces: The Lives and Times of Twelve Canadian Fighter Pilots* (Toronto: Lorimer, 1990), 5.
10. James G. Diehl, "Review of *All the Fine Young Eagles*," *Airpower Journal* (online) (Spring 1998).
11. J. L. Granatstein, *Who Killed the Canadian Military?* (Toronto: Harper, 2004).
12. Everest Riccioni, "A History and Military Analysis of the USAF F-22 Raptor Acquisition," unpublished manuscript, 2005.
13. E-mail from Rolly West, July 21, 2005.
14. Stuart E. Soward, *Hands to Flying Stations*, vol. 2 (Victoria, B.C.: Neptune Developments, 1995), 424–25.
15. Ibid., 246.
16. Ibid., 235–236.
17. Thomas M. Tomlison, *The Threadbare Buzzard: A Marine Fighter Pilot in WWII* (St. Paul, Minn.: Zenith, 2004), 46.
18. Ibid., 45.
19. Soward, *Hands to Flying Stations*, 71.
20. "NATO Tiger Meet Winners," www.natotigers.org/ntmpage/ntm_06.html.
21. Luke Swan, "Great Planes: The F-104 Starfighter," AVI International (Australia), 1989.

22. E-mail from David Bashow, May 1, 2006.
23. Samuel J. Walker, "Interoperability at the Speed of Sound: Canada–United States Aerospace Cooperation . . . Modernizing the CF-18 Hornet," Center for International Relations, Queen's University, Kingston, Ontario, Canada.
24. Terry Johnson, "Four Kills in Five Minutes," *Alberta Report*, November 11, 1996.
25. Ibid.
26. David Bashow et al., "Mission Ready: Canada's Role in the Kosovo Air Campaign," *Canadian Military Journal* (Spring 2000): 56.
27. Ibid., 58.
28. Ibid.
29. Walker, "Interoperability at the Speed of Sound."
30. Tim Hunter et al., "Base Visit #09, Tiger Meet of the Americas 2001," *Sharpshooter 2001*, www.sharpshooter-maj.com/html/bv09.htm.
31. Ibid.
32. Paul Cellucci, "Remarks by Ambassador Paul Cellucci to the American Assembly," February 4, 2005, Arden House, Harriman, New York, available at www.usembassycanada.gov/content/content.asp?section=embconsul&subsection1=cellucci_speeches&document=cellucci_020405.
33. Tom Clancy, *Fighter Wing* (New York: Berkley Books, 1995), 245.
34. "Become a Pilot," www.airforce.forces.gc.ca/pilot/training_e.asp.
35. Jack Dorsey, "Special Report: Training Is Touch and Go," *Virginian-Pilot* (online), September 13, 2004.
36. "Cubic's Advanced Combat Training System Prepares NATO Forces for Combined Operations at Exercise Maple Flag 04," *Business Wire*, June 16, 2004.
37. "A Strong Heritage . . . A Commanding Future," www.airtraining.forces.gc.ca/history_e.asp.
38. "Canadian Military Assumes Command of Afghanistan PRT," American Forces Press Service, August 18, 2005, available at i-newswire.com/pr42708.html.
39. Austin Bay, "'Canadian Military' an Oxymoron?" StrategyPage.com, January 25, 2006.
40. Thomas S. Axworthy, "Strength Needed for Peace," *Legion Magazine* (online) (May/June 2003).
41. Jeffry Brock, *The Thunder and the Sunshine*, vol. 2, *With Many Voices* (Toronto: McClelland and Stewart, 1983), 275.
42. Kalev Sepp, "I Would Fight Alongside Canadians Any Time," *Ottawa Citizen* (online), January 2, 2002.
43. Christopher Cerf and Henry Beard, *The Pentagon Catalog* (New York: Workman, 1986), 5, 8, 14.
44. Julie Foster, "Taxpayers Get $76 Screw-ed," *WorldNetDaily*, February 22, 2000.

45. Tom Abate, "Military Waste under Fire: $1 Trillion Missing," *San Francisco Chronicle* (online), May 18, 2003.
46. Mark Zepezauer, *Take the Rich Off Welfare* (Cambridge, Mass.: South End, 2004), 54–55.
47. Paul Ciotti, "Clinton's War on the Navy," *WorldNetDaily*, February 14, 2000.
48. Ibid.
49. Rowan Scarborough, "Naval Air Is Called Downgraded in Report," *Washington Times*, September 15, 2000.
50. Samuel Katz, *Israel's Air Force* (Osceola, Wis.: Motorbooks, 1991), 54.
51. "Interview with Major Russ 'Crancky' Prechtl," www.f-16.net/interviews_article13.html.
52. Yefim Gordon, *Mikoyan-Gurevich MiG-15* (Hinckley, U.K.: Aerofax, 2001), 65–73.
53. John B. Nichols and Barrett Tillman, *On Yankee Station* (Annapolis, Md.: Naval Institute Press, 2001), 81.
54. George C. Wilson, *Supercarrier* (New York: Berkley Books, 1986), 207–208.
55. Ibid., 155.
56. Ibid.
57. Scott Shuger, *Navy Yes, Navy No*, unpublished manuscript, 1988, 158.
58. Ibid., 55.
59. *Statement of Frank C. Conahan, Director National Security and International Affairs Division before the Subcommittee on Legislation and National Security Committee on Government Operations, House of Representatives, on Readiness of Navy Tactical Air Forces* (Washington, D.C.: GAO, November 2, 1983), 3.
60. Ibid.
61. Lon Nordeen, *Fighters over Israel* (New York: Orion Books, 1990), 55, 62.
62. Shlomo Aloni, *Israeli Mirage and Nesher Aces* (Oxford: Osprey, 2004), 80.
63. Ibid., back cover.
64. Ibid., 10–11.
65. Ibid., 81.
66. Andrew Cockburn, *The Threat: Inside the Soviet Military Machine* (New York: Vintage Books, 1984), 227–28.
67. Ivan Rendall, *Rolling Thunder: Jet Combat from World War II to the Gulf War* (New York: Dell, 1997), 205.
68. Richard P. Hallion, *Storm over Iraq: Air Power and the Gulf War* (Washington, D.C.: Smithsonian, 1992), 27–31.
69. Rendall, *Rolling Thunder*, 179.
70. Ronald Spector. *At War at Sea: Sailors and Naval Combat in the Twentieth Century* (New York: Penguin Books, 2001), 354–55.
71. Cockburn, *The Threat*, 235.
72. Spector, *At War at Sea*, 355.

73. Ibid.
74. Greg Vistica, *Fall from Glory: The Men Who Sank the U.S. Navy* (New York: Simon and Schuster, 1997), 128.
75. Robert K. Wilcox, *Scream of Eagles* (New York: Pocket Books, 2005), foreword, 92.
76. Ibid., 90.
77. Ibid., preface.
78. Ibid., 48.
79. Nichols and Tillman, *On Yankee Station,* 42.
80. Christian G. Appy, *Patriots: The Vietnam War Remembered from All Sides* (New York: Penguin, 2003), 214.
81. Ibid.
82. Robert K. Wilcox, *Wings of Fury* (New York: Pocket Books, 2004), 175.
83. Ibid.
84. James P. Stevenson, *The Pentagon Paradox: The Development of the F-18 Hornet* (Annapolis, Md.: Naval Institute Press, 1993), 34–35.
85. Cliff Gromer, "Ultimate High," *Popular Mechanics* (online) (February 1998).
86. Wilcox, *Wings of Fury,* 174.
87. Ibid., 6–8.
88. Franklin Spinney, *Defense Facts of Life* (Boulder, Colo.: Westview, 1985), 84.
89. Wilcox, *Scream of Eagles,* 116.
90. Ibid., 364.
91. Shuger, *Navy Yes, Navy No,* 161.
92. Ibid., 234.
93. Ibid., 161–62.
94. Ibid., 161.
95. Diego Zampini, "North Vietnamese Aces: Mig-17 and Mig-21 Pilots, Phantom and Thud Killers," www.acepilots.com.
96. Wilson, *Supercarrier,* 208.
97. Rodger Claire, *Raid on the Sun* (New York: Broadway, 2004), 75.
98. Ibid., 91.
99. Soward, *Hands to Flying Stations,* 47.
100. Cockburn, *The Threat,* 421.
101. Jeffrey Ethell, "Pelea a Cuchillo en Chile," *Traduccion Revista Fuerza Aerea* (repr. from *Air Combat* [August 1989]), available at www.fach-extraoficial.com/espanol/f5knife.htm.
102. Ibid.
103. Joan L. Piper, *A Chain of Events* (Dulles, Va.: Brassey's, 2001), 13.
104. John Stillion, *Blunting the Talons: The Impact of Peace Operations Deployments on USAF Fighter Crew Combat Skills,* RGSD-147 (Santa Monica, Calif.: RAND, 1999), 117–18.
105. Everest Riccioni, *Description of Our Failing Defense Acquisition System* (Washington, D.C.: Project on Government Oversight, March 8, 2005), 18.

106. Spinney, *Defense Facts of Life,* 90–91.
107. Riccioni.
108. Piper, *A Chain of Events,* 29.
109. Riccioni.
110. James Fallows, *National Defense* (New York, Random House, 1981), 96.
111. Ibid., 97–98.
112. E-mail from Pierre Rochefort, January 1, 2006.
113. Ethell, "Pelea a Cuchillo en Chile."
114. Lon O. Nordeen, "A Half Century of Jet-fighter Combat: Skill, Determination, and Effective Battle Planning and Tactics Are at Least as Important as Excellent Birds of Prey," *Journal of Electronic Defense* (January 2004).
115. Wilcox, *Wings of Fury,* 134–37.
116. Joe Baugher, "Northrop F-5E Tiger II for the U.S. Navy," available at home.att.net/~jbaugher1/f5_28.html.
117. Robert Coram, *Boyd: The Fighter Pilot Who Changed the Art of War* (New York: Back Bay Books, 2004), 231.
118. Robert DeStasio, "DACT Good . . . Thunderstorm Bad!" *Combat Edge* (March 2003).
119. E-mail from David Bashow, May 15, 2006.
120. Yefim Gordon, *Mikoyan MiG-31* (Hinckley, U.K.: Midland, 2005), 145.
121. Ibid., 140.
122. Ibid. 111.
123. Sharkey Ward, *Sea Harrier over the Falklands* (London: Cassell Military, 2006), 7.
124. Ibid., 93.
125. Ibid., 63–64.
126. Ibid., 67.
127. Ibid., 371.
128. Jaime Hunter, *Sea Harrier: The Last All-British Fighter* (Hinckley, U.K.: Midland, 2005), 64.
129. Jon Lake, "Northrop Grumman F-14 Tomcat," *International Air Power Review* (Winter 2001/2002: 60).
130. Robert Wilcox, *Black Aces High* (New York: St. Martin's Paperbacks, 2002), 13.
131. Bob Kress and Paul Gillcrist, "Battle of the Superfighters: F-14D vs. F/A-18E/F (Two Experts Say the Super Hornet Isn't so Super)," *Flight Journal* (online) (February 2002).
132. Bill Sweetman, "Stacking the Deck: Neither Nimble nor Quick, the Super Hornet Is a Jack-of-All-Trades," *Journal of Electronic Defense* (February 2004).
133. Ibid.
134. Stan Crock, "The (Not So) Super Hornet," *Business Week* (online), December 13, 1999.
135. David L. Robb, *Operation Hollywood: How the Pentagon Shapes and Censors the Movies* (Amherst: N.Y.: Prometheus Books, 2004), 182.

136. Ibid., 133.
137. Ibid., 199.
138. Robert F. Dorr et al., *Korean War Aces* (Oxford: Osprey, 1995), 89.
139. Robert F. Dorr and Warren Thompson, *Korean Air War* (St. Paul, Minn.: Motorbooks International, 2003), 186.
140. Alan Diehl, *Silent Knights: Blowing the Whistle on Military Accidents and Their Cover-Ups* (Washington, D.C.: Brassey's, 2002), 117.
141. Gilles Collingnon, "M88-2: Mission Feedback," *Snecma Magazine* (online) (December 2002).
142. "Rafale Order Finalized," *LeWebmag*, October 12, 2004.
143. Paul Ciotti. "Clinton's War on the Navy," *WorldNetDaily*, February 14, 2000.
144. Ibid.
145. Steve Rowe, "Saving Naval Aviation," Naval Institute *Proceedings* (online) (September 2000).
146. Guy Anderson, "US Defence Budget will equal ROW 'within 12 months,'" *Jane's Defence Industry*, 04 May 2005.
147. Will Daniel, "Navy, Marine Corps Aviation Parts Back Orders Hit All-Time Low," *Navy Supply Corps Newsletter* (January–February 2005).
148. Bob Norris, "U.S.N. or U.S.A.F.?" *Tailhook Association*, www.tailhook.org/USN%20USAF.html.
149. Ibid.
150. Ibid.
151. "Interview with Major Gregory Stroud," www.f-16.net/interviews_article16.html,
152. Ward, *Sea Harrier over the Falklands*, 47.
153. Wilcox, *Scream of Eagles*, foreword.
154. James McCrone, "Top Gun: The Next Generation," *National Guard* (January 1999).
155. Tom Clancy, *Carrier* (New York: Berkley Books, 1999), 65.

CHAPTER 11
1. Michael Isenberg, *Shield of the Republic: The United States Navy in an Era of Cold War and Violent Peace* (New York: St. Martin's, 1993), 1: 496.
2. Gloria Kelly, "Escape Training for Real Life Situations," *Maple Leaf* (online), November 12, 2003.
3. Peter Truscott, *Kursk* (London: Pocket Books, 2003), 36.
4. "Construction Begins for Submarine Escape Trainer," *Navy Newsstand* (online), November 7, 2005, www.news.navy.mil/search/display.asp?story_id=20891.
5. Jack Dorsey, "In the Baltic Sea, 14 Navies Test Their Skills," *Virginian-Pilot*, June 22, 1996.
6. Jean H. Morin and Richard H. Gimblett, *Operation Friction: The Canadian Forces in the Persian Gulf* (Toronto: Dundurn, 1997), 94.
7. James F. Dunnigan and Albert A. Nofi, *Dirty Little Secrets* (New York: Quill, 1990), 175.

8. Tony German, *The Sea Is at Our Gates: The History of the Canadian Navy* (Toronto: McClelland and Stewart, 1990), 240.
9. Raymond Perry, "Why Are Navy COs Getting the Ax?" *DefenseWatch*, March 3, 2004.
10. Ibid.
11. Jeffry Brock, *The Thunder and the Sunshine*, vol. 2, *With Many Voices* (Toronto: McClelland and Stewart, 1983), 71–72.
12. Ibid.
13. Ibid., 73.
14. Ibid., 74.
15. Perry, "Why Are Navy COs Getting the Ax?"
16. Donald Vandergriff, "Army Personnel System Stuck in the Wrong Century," *DefenseWatch*, April 17, 2002.
17. Jeffry Brock, *The Dark Broad Seas*, vol. 1, *With Many Voices* (Toronto: McClelland and Stewart, 1981), 222.
18. John B. Nichols and Barrett Tillman, *On Yankee Station* (Annapolis, Md.: Naval Institute Press, 2001), 44–45.
19. Elmo Zumwalt and [Lt.] Elmo Zumwalt, *My Father, My Son* (New York: Dell, 1987), 42, 44–45, 120.
20. Ibid.
21. Ibid.
22. Kathleen T. Rhem, "Up-or-Out Personnel Policy 'Lousy Idea,' Rumsfeld Tells Sailors," *DEFENSELINK News* (online), November 15, 2003.
23. "U.S. Navy Is Criticized," *Birmingham Post*, February 17, 2001.
24. Alan Diehl, *Silent Knights: Blowing the Whistle on Military Accidents and Their Cover-Ups* (Washington, D.C.: Brassey's, 2002), 304.
25. "Paul Beaver Discusses the U.S. Navy's Public Relations Program Compared with Others around the World," NPR *Weekend Edition: Sunday with Lisa Simeone*, February 17, 2001.
26. E-mail from David Evans, January 26, 2004.
27. Diehl, *Silent Knights*, 184.
28. Scott Shuger, "Why Did the Navy Shoot Down 290 Civilians?" *Washington Monthly*. (October 1988).
29. Ibid.
30. Ibid.
31. Ibid.
32. Alvin Toffler and Heidi Toffler, *War and Anti-War* (New York: Warner Books, 1993), 239.
33. George Friedman and Meredith LeBard, *The Coming War with Japan* (New York: St. Martin's, 1991), 368, 373.
34. Ibid., 359.
35. Charles R. Smith, "Russian 'Rocket' Torpedo Arms Chinese Subs," NewsMax.com, April 24, 2001.
36. Morin and Gimblett, *Operation Friction*, 23.
37. Richard Gimblett, *Operation Apollo: The Golden Age of the Canadian Navy in the War against Terrorism* (Ottawa: Magic Light, 2004), 135.

38. Michael Abrashoff, *It's Your Ship* (New York: Warner Books, 2002), 55–57.
39. E-mail from Aidan Talbott, June 7, 2005.
40. Ibid.
41. Ibid.
42. Ibid.
43. John Chancellor, *Peril and Promise: A Commentary on America* (New York: HarperPerennial, 1991), 29–32.
44. Ibid.
45. Bill Fairburn, "The Canadian Forces Today: The Navy Charts Its Course," *Legion Magazine* (online) (May–June 1999).
46. Department of National Defence, *Report on Canadian Patrol Frigate Cost and Capability Comparison*, 7050-11-11 CRS (Ottawa: Chief Review Services, March 26, 1999), 6–7, C-1, D-1.
47. Ibid.
48. Ibid.
49. Ibid.
50. Ibid.
51. Beryl P. Wajsman, "A Reminder of Our Nation's Pride and Purpose: A Day aboard the HMCS *Montreal*," Institute for Public Affairs of Montreal, September 10, 2004.
52. Ibid.
53. Gimblett, *Operation Apollo*, 50.
54. E-mail from Aidan Talbott, July 12, 2005.
55. Ibid.
56. Abrashoff, *It's Your Ship*, 161.
57. Ibid.
58. Nate Orme, "Experimental Vessel Making Waves for Army," *Flagship* (online), September 18, 2003.
59. Gimblett, *Operation Apollo*, 34.
60. Ibid., 51.
61. See www.snowbirds.dnd.ca/site/faq/faq_e.asp#3 and en.wikipedia.org/wiki/Blue_Angels.
62. Roger Thompson, "Canada's Sea King Ground Crew," *Air Force* (Fall 1993): 5.
63. Ibid.
64. "New Generation Techs Graduate at CFB Borden," *Courier: The Community Newspaper of 4 Wing Cold Lake* (online), July 19, 2005.
65. Jason Hudson, "Naval Service Isn't a Remedial Social Program," Naval Institute *Proceedings* (February 2003): 72–73.
66. Gerald L. Atkinson, *From Trust to Terror: Radical Feminism Is Destroying the U.S. Navy* (Camp Springs, Md.: Atkinson Associates, 1997), 57–58.
67. Kevin O'Neal, "Why I Am Failing My Junior Officers," Naval Institute *Proceedings* (July 2003): 40–42.
68. Ibid.

69. Scott Shuger, *Navy Yes, Navy No*, unpublished manuscript, 1988, 152–53.
70. Isenberg, *Shield of the Republic*, 479.
71. Robert Williscroft, "Lessons from Two Aging Aircraft Carriers," *DefenseWatch*, October 9, 2002.
72. "It's All About Admitting, Learning From, But Not Repeating Mistakes," *Fathom* (October–December 2003): 6–11.
73. Ibid.
74. Ibid.
75. Jack Dorsey, "Navy Made Scapegoat of ex- *Radford* Skipper: Frustrated Officers Say Navy Officials Deny Claims That System Breakdowns Mean Ships Are Sailing with Major Problems," *Virginian-Pilot* (online), July 17, 1999.
76. John F. Schank et al., *Finding the Right Balance: Simulator and Live Training for Navy Units* (Santa Monica, Calif.: RAND, 2002), 25.
77. Franklin C. Spinney, "Trip Report: VFA-81, Cecil Field, 12–15 July 1994," *Defense and the National Interest* (July 1994).
78. Franklin C. Spinney, "Trip Report: Navy and Air Force Air Combat Training," *Defense and the National Interest* (January 30, 2000).
79. Schank et al., *Finding the Right Balance*, 40.
80. Bradley Peniston, *Around the World with the U.S. Navy: A Reporter's Travels* (Annapolis, Md.: Naval Institute Press, 1999), 17.
81. Ronald Spector, *At War at Sea: Sailors and Naval Combat in the Twentieth Century* (New York: Penguin Books, 2001), 382.
82. Ibid.
83. Stuart Soward, *Hands to Flying Stations*, vol. 2 (Victoria, B.C.: Neptune Developments, 1995), 187.
84. Ibid.
85. Raymond Perry, "Why We Almost Lost the Submarine," *DefenseWatch*, April 18, 2005.
86. Ibid.
87. Schank et al., *Finding the Right Balance*, 40.
88. Hugh McManners, *Top Guns* (London: Network Books, 1996), 24.
89. Ibid.
90. Ibid., 177.
91. Richard Gabriel, *Military Incompetence: Why the American Military Doesn't Win* (New York: Noonday, 1985), 15.
92. Sydney Freedberg, "Reforming the Ranks," *National Journal* (online), August 4, 2001.
93. Isenberg, *Shield of the Republic*, 493.
94. Ibid., 415.
95. Scott Shuger, "The Navy We Need and the One We Got," *Washington Monthly* (March 1989).
96. Jon Dougherty, "Admiral Fears U.S. Not Ready," *WorldNetDaily*, September 9, 2000.

97. Schank et al., *Finding the Right Balance,* 24.
98. Spinney, "Trip Report: VFA-81, Cecil Field, 12–15 July 1994."
99. Gordon Peterson, "Training to the Highest Level," *Sea Power* (December 2001).
100. Ibid.
101. Ibid.
102. Daniel S. Appleton, "Sailors in America's Naval Warships Are Not Prepared to Cope with the Violence of Battle," www.dacom.com/fighting-ability/.
103. Ibid.

CHAPTER 12

1. Kevin Keating. "Harass the Brass: Mutiny, Fragging and Desertions in the U.S. Military," *Alternative Press Review* 6, no. 2 (Summer 2001) (online).
2. Terry H. Anderson, *The Movement and the Sixties* (New York: Oxford University Press, 1995), 377.
3. Keating, "Harass the Brass."
4. William F. Arkin and Joshua Handler, *Naval Nuclear Accidents at Sea,* Neptune Papers III: (Washington, D.C.: Greenpeace, 1989), available at www.greenpeace.org/raw/content/international/press/reports/naval-nuclear-accidents.pdf.
5. Gregory A. Freeman, *Sailors to the End: The Deadly Fire on the USS Forrestal and the Heroes Who Fought It* (New York: Perennial, 2002), 2, 8, 10–11, 14.
6. Ronald Spector, *At War at Sea: Sailors and Naval Combat in the Twentieth Century* (New York: Penguin Books, 2001), 367.
7. S. G. Gorshkov, *The Sea Power of the State* (Oxford: Pergamon, 1980), 252.
8. Charles Peters, *How Washington Really Works* (Cambridge: Perseus, 1993), 88.
9. Ibid.
10. Spector, *At War at Sea,* 135.
11. Richard Goldstein, "Elmo R. Zumwalt Jr., Admiral Who Modernized the Navy, Is Dead at 79," *New York Times,* January 3, 2000, A17.
12. John Lehman, *Command of the Seas* (New York: Charles Scribner's Sons, 1988), 97.
13. Michael Abrashoff, *It's Your Ship* (New York: Warner Books, 2002), 169.
14. Jack Spencer, *The Facts about Military Readiness,* Backgrounder 1394 (Washington, D.C.: Heritage Foundation, September 15, 1999), 11–12.
15. Michael Isenberg, *Shield of the Republic: The United States Navy in an Era of Cold War and Violent Peace* (New York: St. Martin's, 1993), 1: 496.
16. Spector, *At War at Sea,* 370.
17. James A. Alcott, *Printed Media in the Year 2001* (Kansas City, Mo.: Midwest Research Institute, 1978), 3.

18. Lehman, *Command of the Seas,* 99.
19. Richard Gabriel, *Military Incompetence: Why the American Military Doesn't Win* (New York: Noonday, 1985), 31.
20. Ibid.
21. Lewis E. Lehrman, *Address to Garrison Forest School* (Owings Mills, Md.: Lehrman Institute: September 23, 1993), 1.
22. Rod Powers, "Minimum Required ASVAB Scores and Education Level," About.com, 2005, usmilitary.about.com/cs/genjoin/a/asvabminimum.htm.
23. Valdas Anelauskas, *Discovering America as It Is* (Atlanta: Clarity, 1999), 167.
24. Ibid.
25. Ibid., 183.
26. Ibid., 181.
27. Ibid., 183.
28. Ibid.
29. "World University Rankings," *Times Higher Education Supplement* (online), November 5, 2004, and "Reflection on Higher Education in the US and UK," *Proceedings of the BRLSI* 5 (2001).
30. Barbara E. Lovitz, "Making the Implicit Explicit: A Conceptual Approach for Assessing the Outcomes of Doctoral Education," paper presented at the ASHE Symposium on Assessing Doctoral Students. Sacramento, California, November 2004.
31. Sun-jung Kim, "A Changed World for Korea's Returnees," *JoongAng Daily* (online), July 27, 2005.
32. Gerald L. Atkinson, "ADM Charles R. Larson: The Anatomy of a Closet Leftie," www.newtotalitarians.com/AnatomyOfAClosetLeftist.html.
33. Ibid.
34. Ibid.
35. Ibid.
36. "Japan: Recruitment and Conditions of Service," *Exploitz.com Travel and Adventure Guide* (online) (January 1994).
37. Anelauskas, *Discovering America as It Is,* 190.
38. H. Yamamoto, "Recent Trends of Drug Abuse in Japan," *Annals of the New York Academy of Sciences*, no. 1025 (2004): 430–38.
39. Spector, *At War at Sea,* 367–68.
40. Sherry Sontag and Christopher Drew, *Blind Man's Bluff* (New York: HarperPaperbacks, 1998), 317.
41. E-mail from Andy Karam, March 22, 2006.
42. Gregory Vistica, *Fall from Glory: The Men Who Sank the U.S. Navy* (New York: Simon and Schuster, 1997), 29.
43. See James W. Crawley, "Navy Charges SEAL Officer with Using Drugs: Seven Other Also Being Investigated," *San Diego Union/Tribune* (online), August 22, 2004, and "Military Sees Drug Use Rise Despite Tests and Warnings," *San Diego Union/Tribune* (online), July 29, 2002.

44. Robert Heinl, "The Collapse of the Armed Forces," *Armed Forces Journal*, June 7, 1971.
45. Paul Ciotti, "Clinton's War on the Navy," *WorldNetDaily*, February 14, 2000.
46. Donald Johnson, *It Wasn't Just a Job: It Was an Adventure* (Lincoln, Neb.: Writers Club, 2002), 141.
47. "Drug Testing: An Overview," ACLU.org, October 21, 2002.
48. Jacob Sullum, *Saying Yes: In Defense of Drug Use* (New York: Tarcher/Putnam: 2003), 117.
49. Doug Thorburn, *How to Spot Hidden Alcoholics: Using Behavioral Clues to Recognize Addiction in Its Early Stages* (Northridge, Calif.: Galt, 2004), 38.
50. "Drug Testing."
51. Anelauskas, *Discovering America as It Is*, 136.
52. Tim Weiner, "U.S. Is No. 1—in Murder, Rape, Robbery," *San Jose Mercury News*, March 13, 1991.
53. Jon Dougherty. "Fat Soldiers Equals Fat Chance on the Battlefield," *WorldNetDaily*, January 3, 2001.
54. Gaetan Lafortune, "Weighty Problem," *OECD Observer* (online) (September 2003).
55. Dougherty, "Fat Soldiers Equals Fat Chance on the Battlefield."
56. "Many Americans Too Fat to Fight: Not Only Are U.S. Troops Out of Shape, Recruits Too Big to Join, Army Says," *Halifax Herald* (online), July 4, 2005.

CHAPTER 13

1. Dale Sykora, "An Operational Perspective of Submarine Evasion Operations," presentation at the 4th International New Horizons in Search Theory Workshop: Investigating Hider Theory. Newport RI, April 2004, available at old.alidade.net/events/040427_SearchTheory/05-submarine-evasion.htm.
2. Scott Shuger, "Paperback Fighter," *Washington Monthly* (November 1989).
3. Ibid.
4. Douglas Waller, *Big Red: The Three Month Voyage of a Trident Nuclear Submarine* (New York: HarperTorch, 2001), 318.
5. Andrew Karam, *Rig Ship for Ultra-Quiet* (Sydney: Sid Harta, 2002). 192–93.
6. Ibid.
7. Waller, *Big Red*, 87.
8. Ronald Spector, *At War at Sea: Sailors and Naval Combat in the Twentieth Century* (New York: Penguin Books, 2001), 336.
9. Patrick Tyler, *Running Critical: The Silent War, Rickover, and General Dynamics* (New York: Harper and Row, 1986), 103.
10. E-mail from Andy Karam, April 13, 2005.

11. Spector, *At War at Sea*, 376.
12. Richard Compton-Hall, *Submarine versus Submarine* (Toronto: Collins, 1988), 32, 107–108, 185.
13. Ibid.
14. Ibid., 107.
15. Scott Shuger, *Navy Yes, Navy No*, unpublished manuscript, 1988, 199–200.
16. Alan J. Snowie, *The Bonnie: HMCS Bonaventure* (Richmond Hill, B.C.: Firefly Books, 1987), 7, 88–89, 230.
17. Ibid.
18. Ravi Sharma, "Engagement on the High Seas," *Frontline* (online), October 25–November 7, 2003.
19. "Navy Orders Complete Standdown," CBS News website, September 15, 2000.
20. Peter Hayes, Lyuba Zarsky, and Walden Bello, *American Lake: Nuclear Peril in the Pacific* (Middlesex: Penguin Books, 1987), 282.
21. William F. Arkin and Joshua Handler, *Naval Nuclear Accidents at Sea*, Neptune Papers III: (Washington, D.C.: Greenpeace, 1989), available at www.greenpeace.org/raw/content/international/press/reports/naval-nuclear-accidents.pdf.
22. Ibid.
23. Shay Cullen. "If the Nuclear Meltdown Happens, What Then?" *Philippine Daily Inquirer*, February 15, 1990.
24. Michael DiMercurio and Michael Benson, *The Complete Idiot's Guide to Submarines* (New York: Alpha. 2003), 70–71.
25. Arkin and Handler, *Naval Nuclear Accidents at Sea*.
26. Tyler, *Running Critical*, 299.
27. DiMercurio and Benson, *The Complete Idiot's Guide to Submarines*, inside cover.
28. Peter Truscott, *Kursk* (London: Pocket Books, 2003), 57, 130.
29. David Kaplan, "When Incidents Are Accidents: The Silent Saga of the Nuclear Navy," *Oceans Magazine* (August 1983).
30. Investigative Reports (with Bill Kurtis), "*Out of the Gulf, Into the New Navy.*" A&E Entertainment, 2000; and "Questions Remain after Canadian Navy Divers' Deaths," *Wednesday Report*, February 13, 1991.
31. Scott Shuger, "*Fall from Glory: The Men Who Sank the U.S. Navy*, Book Review," *Washington Monthly* (May 1996).
32. Alan Diehl, *Silent Knights: Blowing the Whistle on Military Accidents and Their Cover-Ups* (Washington, D.C.: Brassey's, 2002), 32.
33. Jaya Tiwari and Cleve J. Gray, "U.S. Nuclear Weapons Accidents," Center for Defense Information, n.d., available at www.cdi.org/Issues/NukeAccidents/accidents.htm.
34. Ibid.
35. Ibid.

36. Hayes et al., *American Lake*, 287.
37. "U.S. Nuclear Ships May Conceal Accidents," *Green Left Weekly Online Edition*, no. 52, April 15, 1992.
38. Hayes et al., *American Lake*, 267.

CHAPTER 14
1. Peter Cary, "Death at Sea," *U.S. News & World Report*, April 23, 1990.
2. Henry Scammell, *Giantkillers: The Team and the Law That Help Whistleblowers Recover America's Stolen Billions* (New York: Atlantic Monthly, 2004), 45.
3. Ibid.
4. Scott Shuger, *Navy Yes, Navy No*, unpublished manuscript, 1988, 167.
5. Tim Weiner, "Arms Fiascoes Lead to Alarm inside Pentagon," *New York Times on the Web: Learning Network (Student Connections)*, June 8, 2005.
6. Hedrick Smith, *The Power Game* (New York: Ballantine Books, 1988), 167.
7. Citizens against Government Waste, *The Pig Book: How Government Wastes Your Money* (New York: Thomas Dunne Books, 2005), 38–39.
8. Lauren Holland, *Weapons under Fire* (New York: Garland, 1997), 161.
9. Samuel P. Huntington, *The Soldier and the State* (Cambridge, Mass.: Belknap, 2003), 266.
10. Valdas Anelauskas, *Discovering America as It Is* (Atlanta: Clarity, 1999), 168.

CHAPTER 15
1. "Edmund Burke Quotes and Quotations Compiled by GIGA," www.giga-usa.com/quotes/authors/edmund_burke_a001.htm.
2. David Adams, "We are Not Invincible," Naval Institute *Proceedings* (May 1997): 35–39.
3. "Dwight D. Eisenhower Quotes," www.brainyquote.com/quotes/quotes/d/dwightdei149090.html.

Bibliography

ARTICLES (PRINT AND ONLINE)
Abate, Tom. "Military Waste under Fire: $1 Trillion Missing." *San Francisco Chronicle* (online), May 18, 2003.
———. "War Game Reveals Navy Risk." *San Francisco Chronicle* (online), March 20, 2003.
Aboulafia, Richard. "Maritime Patrol Market: Escaping the Doldrums." *Aerospace America* (online), April 2002.
"About Dick Marcinko." www.dickmarcinko.com/AboutDM.aspx
Adams, David. "We Are Not Invincible." Naval Institute *Proceedings*, May 1997.
"Aircraft Carrier Article Classified Secret by Navy." *Navy Times*, May 24, 1982.
Alexander, Greg. "Iranian Air Force F-14." Aerospace.org, May 12, 2002.
Aloni, Shlomo. "Israel's Roving Warriors." *Air Forces Monthly*, no. 5 (1999).
Anderson, Guy. "U.S. Defence Budget Will Equal ROW Combined 'within 12 Months.'" *Jane's Defence Industry* (online), May 4, 2005.
Anderson, Rick. "Peril Harbor." *Seattle Weekly* (online), January 11–16, 2001.
———. "Sub Slasher Sentenced." *Seattle Weekly* (online), August 2–8, 2001.
Atkinson, Gerald L. "ADM Charles R. Larson: The Anatomy of a Closet Leftie." *New Totalitarians* website, www.newtotalitarians.com/AnatomyOfAClosetLeftist.html.
"Aussie Duo Conquer Perisher Challenge." *Navy News*, November 2002. Republished by Dutchsubmarines.com, www.dutchsubmarines.com/specials/special_aussie_duo.htm.
Axworthy, Thomas S. "Strength Needed for Peace." *Legion Magazine* (online) (May/June 2003).

"Band-Aid Navy: How Shortages Are Burning Out Sailors and Wearing Out the Fleet." *Navy Times*, May 22, 2000.

Bashow, David, et. al. "Mission Ready: Canada's Role in the Kosovo Air Campaign." *Canadian Military Journal* (Spring 2000).

Bay, Austin. "'Canadian Military' an Oxymoron?" StrategyPage.com, January 25, 2006.

Beheim, Eric. "New Command to Integrate Navy's ASW Mission." *All Hands* (August 2004).

Bellis, Mary. "Inventions of France: French Inventors." About.com.

Benedict, John R. "The Unraveling and Revitalization of U.S. Navy Antisubmarine Warfare." *Naval War College Review* 58, no. 2 (Spring 2005).

Bennington, Ashley. "Stealthy Subs (letter)." *Popular Science*, July 1, 1999.

Borger, Julian. "Uncle Sam Sunk in Credibility Gulf." *The Age* (online), September 8, 2002.

Bowman, Martin. "Snowbirds!" *Air Classics* (January 2001).

Brattebo, Douglas M. "From the Guest Editor." *White House Studies* (Spring 2004).

Brewster, Murray. "Other Countries Likely First to Use Canadian Mine-Clearing Drones." Canada.com, July 21, 2005.

Brooke, James. "U.S. Rule of Pacific Waves Faces China Challenge." *International Herald Tribune* (online), December 30, 2004.

Caldwell, Robert J. "Navy's Woes Reflect Risks of Years of Underfunding Defense." *San Diego Union Tribune* (online), September 24, 2000.

"Canada's Air Force: World War I." *Canadian Air Force*, www.airforce.forces.gc.ca/hist/ww_1_e.asp.

"Canadian Military Assumes Command of Afghanistan PRT." American Forces Press Service, August 18, 2005, available at i-newswire.com/pr42708.html.

Carpenter, Ted Galen, and Justin Logan. "Eastern Fronts." *American Prospect Online Edition*, May 16, 2005, www.prospect.org/web/page.ww?section=root&name=ViewWeb&articleId=9670.

"Carrier *Kennedy* Skipper Relieved of Command: Ship Failed Major [Navy Board] INSURV Inspection." *Navy Times*, December 17, 2001.

Cary, Peter. "Death at Sea" *U.S. News & World Report*, April 23, 1990.

———. et al. "What's Wrong with the Navy?" *U.S. News & World Report*, July 13, 1992.

"CDC's CANTASS is an Exceptional Canadian Achievement." *Wednesday Report* 4, no. 36 (online).

Chernitzer, Rick. "Problems Corrected, USS John F. Kennedy Ready to Go." *Stars and Stripes* (online), February 25, 2002.

Cloutier, Todd. "Daring to Go Dutch." *Undersea Warfare Magazine* (online) (Fall 2003).

Ciotti, Paul. "Clinton's War on the Navy." *WorldNetDaily,* February 14, 2000.

Collingnon, Gilles. "M88-2: Mission Feedback." *Snecma Magazine* (online) (December 2002).

Conley, Daniel. "Don't Discount the Diesel." Naval Institute *Proceedings* (October 1987).

"Construction Begins for Submarine Escape Trainer." *Navy Newsstand* (online), November 7, 2005, www.news.navy.mil/search/display.asp?story_id=20891.

Corpus, Victor N. "If It Comes to a Shooting War." *Asia Times Online*, April 20, 2006.

Coyle, Frank G. "Navy Needs Heavy-Lift/Countermine Helos." Naval Institute *Proceedings* (August 2004).

Crawley, James W "Military Sees Drug Use Rise Despite Tests and Warnings." *San Diego Union/Tribune* (online), July 29, 2002.

———. "Navy Charges SEAL Officer With Using Drugs: Seven Other Also Being Investigated." *San Diego Union/Tribune* (online), August 22, 2004.

———. "S.D. might be Home to Swedish Sub." *San Diego Union/Tribune* (online), October 14, 2004.

———. "Quiet Diesel Subs Surface as New Threat." *San Diego Union/Tribune* (online), January 22, 2004.

———. "Vikings' Funeral: Two S-3B Jet Squadrons Being Retired at North Island Naval Air Station." *San Diego Union/Tribune* (online), April 16, 2004.

Crock, Stan. "The (Not So) Super Hornet." *Business Week* (online), December 13, 1999.

"Cubic's Advanced Combat Training System Prepares NATO Forces for Combined Operations at Exercise Maple Flag 04." *Business Wire,* June 16, 2004.

Cullen, Shay. "If the Nuclear Meltdown Happens, What Then?" *Philippine Daily Inquirer,* February 15, 1990.

Daniel, Will. "Navy, Marine Corps Aviation Parts Back Orders Hit All-Time Low." *Navy Supply Corps Newsletter* (January–February 2005).

Day, Vox. "No Case for Internment." *WorldNetDaily,* September 6, 2004.

"Defense Watch." *Defense Daily,* June 18, 2001.

Delgado, James "Diving at Ground Zero: The Ships of Bikini Atoll Are a Grim Reminder of the Atomic Age," available at www.jamesdelgado.com/Diving_at_Ground_Zero.pdf.

DeStasio, Robert. "DACT Good . . . Thunderstorm Bad!" *Combat Edge* (March 2003).

Diehl, James G. "Review of *All the Fine Young Eagles.*" *Airpower Journal* (online) (Spring 1998).

Digges, Charles. "Italy Furious after U.S. Navy Tried to Cover Up Sub Accident." *Bellona News* (online), November 11, 2003.

Director, Navy Equal Opportunity Office (PERS-670). "Navy-Wide Demographic Data Report for Q4 2004." Millington, Tenn., November 8, 2004.

Dixon, Robyn, and Paul Richter. "Russia Brags Its Jets Buzzed Carrier *Kitty Hawk.*" *Virginian Pilot,* November 16, 2000.

Dolan, Matthew. "Increasing Number of Navy Officers Getting Fired." *Virginian-Pilot,* March 10, 2004.

Dombrowski, Peter J., and Andrew L. Ross. "Transforming the Navy: Punching a Feather Bed?" *Naval War College Review* 56, no. 3 (Summer 2003).

Donahoe, C. A. "Mines: Will They Sink the U.S. Navy?" Marine Corps University Command and Staff College, 1992, available at www.globalsecurity.org/military/library/report,1992/DCA.htm.

Dorsey, Jack. "Ex-SEALS Say Racism Rampant." *Virginian-Pilot,* July 15, 2001.

———. "In the Baltic Sea, 14 Navies Test Their Skills." *Virginian-Pilot,* June 22, 1996.

———. "Navy Made Scapegoat of ex- *Radford* Skipper: Frustrated Officers Say Navy Officials Deny Claims That System Breakdowns Mean Ships Are Sailing with Major Problems." *Virginian-Pilot* (online), July 17, 1999.

———. "Special Report: Training Is Touch and Go." *Virginian-Pilot* (online), September 13, 2004.

———. and Dale, Eisman. "Report Spotlights Ship's Problems." *Virginian-Pilot* (online), July 13, 2005.

Dougherty, Jon E. "Admiral Fears U.S. Not Ready." *WorldNetDaily,* September 9, 2000.

———. "China Simulates Attacks on U.S. Carriers." *WorldNetDaily,* December 18, 2001.

———. "China to Target U.S. Carriers." *WorldNetDaily,* July 13, 2002.

———. "Fat Soldiers Equals Fat Chance on the Battlefield." *WorldNetDaily*, January 3, 2001.
———. "Navy, Air Force Spare-Parts Crisis." *WorldNetDaily*, September 30, 2000.
———. "Navy Combat Systems Unsafe? Chief Tester Says Weapons Deployed before 'Acceptable.'" *WorldNetDaily*, October 1, 2002.
———. "Pakistan to Put Nukes on Subs." *WorldNetDaily*, February 23, 2001.
———. "Russian Flyover Takes Navy by Surprise." *WorldNetDaily*, December 7, 2000.
———. "Why Is the Navy Home?" *WorldNetDaily*, December 28, 1999.
Draper, Robert. "Reflections on Higher Education in the U.S. and the U.K." *Proceedings of the BRLSI* (online) 5 (2001).
"Drug Testing: An Overview." ACLU.org, October 21, 2002.
"Duct-Tape Aviation: How Old Planes and Parts Shortages Are Running Aircraft and Sailors Ragged." *Navy Times*, September 10, 2001.
Eisman, Dale. "Four Russian Planes Fly by U.S. Carrier, Send Photos of the Trip." *Virginian-Pilot*, December 8, 2000.
———. "Military Surprised, Disappointed by Results of Race Survey." *Virginian-Pilot*, Novemvber 27, 1999.
"The Era of the Aircraft Carrier May Be Ending." *Honolulu Star-Bulletin*, August 26, 1997.
Ethell, Jeffrey. "Pelea a Cuchillo en Chile." *Traduccion Revista Fuerza Aerea*, available at www.fach-extraoficial.com/espanol/f5knife.htm. (A Spanish translation of an article by Ethell that originally appeared in the August 1989 *Air Combat*.)
Foster, Julie. "Taxpayers Get $76 Screw-ed." *WorldNetDaily*, February 22, 2000.
"F-16 Fighting Falcon History." GlobalSecurity.org, February 23, 2003.
"Facts That May Surprise You." www.canadianembassy.org/ca/facts-mil-en.asp
Fairburn, Bill. "The Canadian Forces Today: The Navy Charts Its Course." *Legion Magazine* (online) (May–June 1999).
Farah, James. "Anchors Away: The Late Great U.S. Navy." *WorldNetDaily*, January 29, 1999.
Fisher, Richard. "Developing U.S.-Chinese Nuclear Naval Competition in Asia." *International Assessment and Strategy Center* website, January 16, 2005.
Freedberg, Sydney J. "Reforming the Ranks." *National Journal* (online), August 4, 2001.

Garran, Robert. "A Deadly Exercise in Stealth." *Weekend Australian,* December 23–25, 2000.

Gertz, Bill. "China Sub Stalked U.S. fleet." *Washington Times,* November 13, 2006.

———. "China Tests Potent, Supersonic Anti-Ship Cruise Missiles." *Washington Times,* September 25, 2001.

———. "Russian Sub Stalks Three U.S. Carriers." *Washington Times,* November 23, 1997.

Goldstein, Lyle, and William Murray. "Undersea Dragons: China's Maturing Submarine Force." *International Security* 28, no.4 (Spring 2004).

———. and Yuri M. Zhukov. "A Tale of Two Fleets: A Russian Perspective on the 1973 Naval Standoff in the Mediterranean." *Naval War College Review* 57, no. 2 (Spring 2004).

———. "Superpower Showdown in the Mediterranean, 1973." *Sea Power* (online) (October 2003).

Goldstein, Richard. "Elmo R. Zumwalt Jr., Admiral Who Modernized the Navy, Is Dead at 79." *New York Times,* January 3, 2000.

Gordon, Robert. "Navy's Ships Making Valuable Contribution." *Halifax Chronicle Herald,* March 23, 1983.

Granatstein, J. L. "The American Influence on the Canadian Military, 1939–1963." *Canadian Military History* (online) (Spring 1993).

Grant, Rebecca. "The Carrier Myth." *Air Force* (online) (March 1999).

Gromer, Cliff. "Ultimate High." *Popular Mechanics* (online) (February 1998).

Healy, Patrick. "Harvard's Quiet Secret: Rampant Grade Inflation." *Boston Globe* (online), October 7, 2001.

Heinl, Robert D. "The Collapse of the Armed Forces." *Armed Forces Journal,* June 7, 1971.

Hellman, Christopher. "Last of the Big Time Spenders: U.S. Military Budget Still the World's Largest, and Growing." *Center for Defense Information,* February 3, 2003, www.cdi.org/issues/wme/spendersFY04.html.

Helms, Nat, and Ray Perry. "Russian Submarine Emergency Sheds Light on U.S. Navy Deep Water Rescue Capabilities." *DefenseWatch,* August 13, 2005.

Hodge, Nathan. "Australian 'Hit' on U.S. Sub Gets Attention." *Defense Weekly Daily Update,* 1 October 2003.

Hoffman, Carl. "Rotary Club." *Air & Space/Smithsonian* (January 2006).

Hoffman, Frank. "Thinking Ahead Intelligently," review of *Uncovering Ways of War: U.S. Intelligence and Foreign Military Innovation,* by Thomas C. Mahnken, *Naval War College Review* 56, no. 1 (Winter 2003): 157.

Holzer, Robert. "Dangerous Waters: Submarines, New Mines Imperil Ill-Prepared U.S. Navy Fleet." *Defense News,* May 4–10, 1998. Quoted in Shirley A. Kan et al., "China's Foreign Conventional Arms Acquisitions: Background and Analysis," *Congressional Research Service Report,* October 10, 2000.

Hooper, Barrett. "How to Fight Like a Man." *Saturday Post,* January 26, 2002.

Hsu, Brian. "Navy Allows a Rare Glimpse of Sub." *Taipei Times,* June 23, 2000, available at www.dutchsubmarines.com/specials/special_glimpse_seatiger.htm.

Hudson, Jason. "Naval Service Isn't a Remedial Social Program." U.S. Naval Institute *Proceedings* (February 2003).

Hunter, Tim, et al. "Base Visit #09, Tiger Meet of the Americas 2001." *Sharpshooter* 2001, available at www.sharpshooter-maj.com/html/bv09.htm.

"IG Issues Scorching Report on Naval Aviation." *Navy Times,* September 25, 2000.

"Iroquois (Tribal) Class," Hazegray.org.

"It's All About Admitting, Learning From, But Not Repeating Mistakes." *Fathom* (October–December 2003).

"Japan: Recruitment and Conditions of Service." *Exploitz.com Travel and Adventure Guide* (online) (January 1994).

Jean, Grace. "Surface Combatants Dominate Future Fleet." *National Defense* (online) (April 2006).

Johnson, Ray, and Lem Robson. "William Tell '96." *Airman* (online) (February 1997).

Johnson, Terry. "Four Kills in Five Minutes." *Alberta Report,* November 11, 1996.

Kaplan, David B. "When Incidents Are Accidents: The Silent Saga of the Nuclear Navy." *Oceans Magazine* (August 1983). (Note: This article is inaccurate in some aspects, but much of it is reasonable and true.)

Kaylor, Robert. "Deadly Games of Hide and Seek." *U.S. News & World Report,* June 15, 1987.

———. "The Navy's 21st-Century Submarine: Critics Ask Whether the *Seawolf* Can Keep the U.S. ahead of the Soviets." *U.S. News & World Report,* April 24, 1989.

Keating, Kevin. "Harass the Brass: Mutiny, Fragging and Desertions in the U.S. Military." *Alternative Press Review* 6, no. 2 (Summer 2001) (online).
Kelly, Gloria. "Escape Training for Real Life Situations." *Maple Leaf* (online), November 12, 2003.
Kelton, Maryanne. "New Depths in Australia-U.S. Relations: The *Collins* Class Submarine Project." School of Political and International Studies Working Paper, Flinders University of South Australia, March 2004.
Kim, Sun-jung. "A Changed World for Korea's Returnees." *JoongAng Daily* (online), 27 July 2005.
Knuth, Dean. "Battle Group Tactical Proficiency." Naval Institute *Proceedings* (December 1981).
———. "Do We Still Need a Navy?" *Washington Post*, May 7, 1982.
Kress, Bob, and Paul Gillcrist. "Battle of the Superfighters: F-14D vs. F/A-18E/F (Two Experts Say the Super Hornet Isn't So Super)." *Flight Journal* (online) (February 2002).
Lafortune, Gaetan. "Weighty Problem." *OECD Observer* (online) (September 2003).
Lake, Jon. "Northrop Grumman F-14 Tomcat." *International Air Power Review* (Winter 2001/2002.
Lambie, Chris. "Fast Torpedoes Pose New Threat." *Daily News* (online), August 6, 2004.
Levy, James. "Race for the Decisive Weapon: British, American, and Japanese Carrier Fleets, 1942–1943." *Naval War College Review* 58, no. 1 (Winter 2005).
"A Long Flight to Nowhere." *Citizens against Government Waste* website, February 18, 2005.
Lumpkin, John J. "Collision with Carrier Raises Concerns." *Union Leader* (online), August 6, 2004.
———. "Navy Relieves 80 Commanders over 5 Years." *Guardian* (online), December 22, 2004. (This article, although published in a British newspaper, concerns the U.S. Navy.)
"Many Americans Too Fat to Fight: Not Only Are U.S. Troops Out of Shape, Recruits Too Big to Join, Army Says." *Halifax Herald* (online), July 4, 2005.
Margolis, Eric. "Did Russia Win D-Day?" *CNEWS*, May 30, 2004.
Marolda, Ed. "Mine Warfare." *Naval Historical Center* website, available at www.history.navy.mil/wars/korea/minewar.htm.
Mayfield, Dave. "Subs Emerge from the Deep." *Virginian-Pilot*, April 27, 1998.

McCrone, James. "Top Gun: The Next Generation." *National Guard* (January 1999).
McPhedran, Ian. "*Collins* Subs a Lethal Force." News.com.au, September 24, 2003.
McMichael, William H. "Today's Military Integrated, but Racism Still Exists." *Newport News Daily Press,* July 24, 1998.
Mendenhall, Corwin. "The Case for Diesel Submarines." Letter to Naval Institute *Proceedings* (September 1995), 27.
Mintz, Morton. "Article Critical of Carriers Stamped 'Secret' by Navy." *Washington Post,* May 4, 1982.
Mooney, Shane. "The A-Teams." *Maxim* (March 2001).
Moore, Mike. "The Able-Baker-Where's Charlie Follies: The Story of Operation Crossroads Is a Sad Tale of U.S. Naval Arrogance . . . and Ignorance." *Bulletin of the Atomic Scientists,* May 1, 1984.
Morley, Mike. "*Kitty Hawk* Packs a Punch in Tandem Thrust '99." COMVAIRPAC Public Affairs Office Report, March 30, 1999. (Incidentally, despite the highly suggestive title, the actual report contains no evidence of a "victory" for the *Kitty Hawk*.)
"Naval Aviation Conducts Safety Stand Down to Focus on Safety." *Navy Newsstand,* March 3, 2006.
"Navy Orders Complete Standdown." CBS News website, September 15, 2000.
"Navy to Get Early Crack at Mine-Hunting System." *Halifax Herald* (online), July 19. 2005.
Nedelchev, Nicholay. "The U.S. Silent Service in Early WWII: A Story of Failure," available at uboat.net/allies/technical/torpedo_problems.htm.
"New Generation Techs Graduate at CFB Borden." *Courier: The Community Newspaper of 4 Wing Cold Lake* (online), 19 July 2005.
Newman, Richard J. "Breaking the Surface." *U.S. News & World Report,* April 6, 1998.
Nicholson, Brendan. "Collins Sub Shines in U.S. War Game." *Sunday Age* (online), October 13, 2002.
———. "Collins Subs Star in Naval Exercises." *The Age* (online), September 24, 2003.
———. "Sub Troops Get Underwater Cover." *The Age* (online), March 2, 2003.
Noel, Martin A. "Review of Knights of the Air: Canadian Fighter Pilots in the First World War." *Air & Space Power Journal* (Fall 2003).

Nordeen, Lon O. "A Half Century of Jet-Fighter Combat: Skill, Determination, and Effective Battle Planning and Tactics Are at Least as Important as Excellent Birds of Prey." *Journal of Electronic Defense* (January 2004).
Norris, Bob. "U.S.N. or U.S.A.F.?" *Tailhook Association*, 2004, www.tailhook.org/USN%20USAF.html.
"Nuclear Submarine Crew Member Charged with Sabotage." *Seattle Post-Intelligencer* (online), January 11, 2001.
Nunez, Joseph. "Canada's Global Role: A Strategic Assessment of Its Military Power." *Parameters*, September 22, 2004.
O'Hanlon, Michael, Lyle Goldstein, and William Murray. "Damn the Torpedoes: Debating Possible U.S. Navy Losses in a Taiwan Scenario." *International Security* 29, no. 2 (Fall 2004).
"O'Neal, Kevin. "Why I Am Failing My Junior Officers." Naval Institute *Proceedings* (July 2003).
"On the Defensive." *Bulletin of the Atomic Scientists* (January 2004).
"Operation Mainbrace." *Time*, September 22, 1952.
Orem, John H. "Security Needs a Professional Officer Community." Naval Institute *Proceedings* (September 2004).
Orme, Nate. "Experimental Vessel Making Waves for Army." *Flagship* (online), September 18, 2003.
"Our Sub Won Attack on U.S. Vessel: Blais." *Toronto Star*, May 9, 1984.
O'Sullivan, Arieh. "Report: IAF Whips U.S. Pilots in Exercise." *Jerusalem Post* (online), September 24, 1999.
Paige, Sean. "GAO Offers Stark Assessment of Warship Vulnerability." *Insight on the News*, August 14, 2000.
Pash, Gerry. "Rim of the Pacific 2004: Aurora Hunts U.S.N Submarine." *Lookout*, July 26, 2004.
Perry, Raymond. "USS *San Francisco:* The Final Verdict" *DefenseWatch*, June 3, 2005.
——. "Why Are Navy COs Getting the Ax?" *DefenseWatch*, March 3, 2004.
——. "Why We Almost Lost the Submarine." *DefenseWatch*, April 18, 2005.
Peterson, Gordon I. "Training to the Highest Level." *Sea Power* (December 2001).
Philip, David. "Victory in Sight for Cold Lake." *Alberta Report*, May 27, 1996.
Powers, Rod. "Minimum Required ASVAB Scores and Education Level." About.com, 2005, usmilitary.about.com/cs/genjoin/a/asvabminimum.htm.

Powis, Jonathan. "U.K.'s *Upholder*-Class Subs Go to Canada." Naval Institute *Proceedings* (October 2002).

Priestly, S. T. "Politics, Procurement Practices, and Procrastination: The Quarter-Century *Sea King* Helicopter Replacement Saga." *Canadian American Strategic Review* (online) (January 2004).

"Questions Remain after Canadian Navy Divers' Deaths." *Wednesday Report*, February 13, 1991.

"Rafale Order Finalized." *LeWebmag*, October 12, 2004.

Rhem, Kathleen T. "Up-or-Out Personnel Policy 'Lousy Idea,' Rumsfeld Tells Sailors." *DEFENSELINK News* (online), November 15, 2003.

Ricks, Thomas E. "Drugs, Sex and Recommendations . . ." *Washington Post*, July 17, 2001.

Rivers, Brendan "Forward . . . into Dangerous Waters." *Journal of Electronic Defense* (October 1999).

Rogers, R. W. "Naval Commander Says Drugs, Lack of Money Contribute to Fleet's Problems." *Daily Press*, July 27, 2001.

Rowe, Steve. "Saving Naval Aviation." Naval Institute *Proceedings* (online) (September 2000).

Russell Chaddock, Gail. "U.S. Notches World's Highest Incarceration Rate." *Christian Science Monitor* (online), August 18, 2003.

"Russia Promises Weapons to China to Destroy U.S. Navy." NewsMax.com, November 16, 2000.

Ryan, Paul. "LCS Will Transform Mine Warfare." Naval Institute *Proceedings* (December 2004).

Sarty, Roger. "The Royal Canadian Navy and the Battle of the Atlantic, 1939–1945." Canadian War Museum, Ottawa, 2001, available at www.civilization.ca/cwm/disp/dis007_e.html.

Scarborough, Rowan. "Naval Air Is Called Downgraded in Report." *Washington Times*, September 15, 2000.

———. "Navy Officer Gives Readiness Warning: Says His Planes Are Not Fit to Fight." *Washington Times*, May 3, 1999.

———. "Skill Level of Pilots in Air Force, Navy Seen as Degraded." *Washington Times*, February 21, 2000.

———. "U.S. Ship Took 40 Minutes to Respond to Order." *Washington Times*, December 7, 2000.

Sepp, Kalev. "I Would Fight Alongside Canadians Any Time." *Ottawa Citizen* (online), January 2, 2002.

Sharma, Ravi. "Engagement on the High Seas." *Frontline* (online), October 25–November 7, 2003.

Shuger, Scott. "Fall from Glory: The Men Who Sank the U.S. Navy, Book Review," *Washington Monthly* (May 1996).
———. "Hurts So Good: At POW School the Navy Locked Me in a Box" *Washington Monthly* (May 1988).
———. "The Navy We Need and the One We Got." *Washington Monthly* (March 1989).
———. "The Navy's Plane Stupidity." *Washington Monthly* (October 1985).
———. "Paperback Fighter." *Washington Monthly* (November 1989).
———. "Why Did the Navy Shoot Down 290 Civilians?" *Washington Monthly* (October 1988).
Shukovsky, Paul. "Ex-Navy Pilot Sent to Prison." *Seattle Post-Intelligencer*, June 15, 2004.
"SMCC 2004: Impressions of a U.S.N Student." Dutchsubmarines.com, www.dutchsubmarines.com/specials/special_smcc_impressions.htm.
Smith, Charles. "Chinese Observers aboard *Kursk*?" *WorldNetDaily*, September 17, 2000.
Spinney, Franklin C. "Trip Report: Navy and Air Force Air Combat Training." *Defense and the National Interest* (January 30, 2000).
———. "Trip Report: VFA-81, Cecil Field, 12–15 July 1994." *Defense and the National Interest* (July 1994).
Strasser, Steven, et al. "Are Big Warships Doomed?" *Newsweek*, 17 May 1982, cited in Roger W. Barnett, "Naval Power for a New American Century," *Naval War College Review* 55, no. 1 (Winter 2002).
Stanton, Jim. "Boeing Lands £2 Billion Deal for Sub Hunters." *Edinburgh Evening News* (online), June 16, 2004.
"Svensk ubat gackar marinen I U.S.A." *Gefle Dagblad* (online), 5 October 2005.
Sweetman, Bill. "Stacking the Deck: Neither Nimble nor Quick, the Super Hornet Is a Jack-of-All-Trades." *Journal of Electronic Defense* (February 2004).
Tanabe, Yahachi, and Joseph D. Harrington. "I Sank the *Yorktown* at Midway." Naval Institute *Proceedings* (May 1963).
"Tandem Thrust '99." *Asia-Pacific Defense Forum* (online) (Fall 1999).
Taplin, Jennifer. "Flood of Memories about Go-Boat Sub." *Halifax Daily News* (online), March 20, 2006.
Tate, Rob. "Malta Spitfire: The Diary of a Fighter Pilot." Book review. *Air & Space Power Journal* (online) (Spring 2004).
"Team Canada Soars to Victory at William Tell '96." *Airman* (online) (December 1996).

Thompson, Charles C., and Tony Hays. "Gore Brings Back $640 Toilet Seat." *WorldNetDaily,* October 27, 2000.
Thompson, Roger. "Are the Nuclear Submarine's Days of Undersea Predominance Numbered?" *International Insights* 6, no. 2 (1990).
———. "Canada's Sea King Ground Crew." *Air Force* (Fall 1993).
Tiwari, Jaya, and Cleve J. Gray. "U.S. Nuclear Weapons Accidents." Center for Defense Information, n.d., available at www.cdi.org/Issues/NukeAccidents/accidents.htm.
Toyka, Viktor. "Ask Questions about Our ability to Conduct Antisubmarine Warfare." Naval Institute *Proceedings* (September 2004).
Turner, Stansfield. "Is the U.S. Navy Being Marginalized?" *Naval War College Review* 56, no. 3 (Summer 2003).
Tyler, Greg. "USS *Vincennes* Cracking Down on Use of Illegal Drugs." *Stars and Stripes* (online), November 20, 2004.
"U.S. Flotilla Decked by Canadian Sub." *Winnipeg Free Press,* May 9, 1984.
"U.S. Navy Is Criticized." *Birmingham Post,* February 17, 2001.
"U.S. Nuclear Ships May Conceal Accidents." *Green Left Weekly Online Edition,* no. 52, April 15, 1992.
"U.S. Pacific Fleet No Match for the Russian Air Force." *Discerning the Times Digest and Newsbytes* (November 2000).
Vandergriff, Donald E. "Army Personnel System Stuck in the Wrong Century." *DefenseWatch,* April 17, 2002.
———. Review of *Not a Good Day to Die: The Untold Story of Operation Anaconda,* www.d-n-i.net/vandergriff/naylor_review.htm.
"Vice Admiral C. A. Lockwood, ComSubPac." Dutchsubmarines.com, www.dutchsubmarines.com/specials/special_lockwood.htm.
Voronin, G. "The Silence of Our Submarines Annoys Not Only Dilettantes." *Krasnaya zvezda,* 28 January 1994.
Wajsman, Beryl. "A Reminder of Our Nation's Pride and Purpose: A Day aboard the HMCS *Montreal.*" Institute for Public Affairs of Montreal, September 10, 2004. (I can do without the jingoism, but there is some interesting material in this short article on the Canadian navy.)
"Wake-Up Call." *Guardian,* September 6, 2002, available at www.guardian.co.uk/g2/story/0,3604,786992,00.html.
Walker, Larry. "Too Expensive to Lose, Too Expensive to Use: Weapons Systems." *Insight on the News,* January 23, 1995.
Wallace, James. "Boeing 737 May Be Enlisted as a Warplane." *Seattle Post-Intelligencer* (online), May 13, 2002.

Waller, J. Michael. "Russian Sub Stalks Three Carriers, Practices Attack on USS *Carl Vinson*." *Russia Reform Monitor* (online), no. 347, December 1, 1997.

Weiner, Tim. "Arms Fiascoes Lead to Alarm inside Pentagon." *New York Times on the Web: Learning Network (Student Connections)*, June 8, 2005.

———. "U.S. Is no. 1—in Murder, Rape, Robbery." *San Jose Mercury News*, March 13, 1991.

Westerman, Toby. "Moscow Selling Aircraft Carrier Killers." *WorldNetDaily*, August 5, 2000.

———. "Naval Officer Warns of Attack: Daly Says Nuclear Sub, Carrier Bases Vulnerable." *WorldNetDaily*, June 19, 2001.

Williscroft, Robert. "Is the Nuclear Submarine Really Invincible?" *DefenseWatch*, September 9, 2004.

———. "Lessons from Two Aging Aircraft Carriers" *DefenseWatch*, October 9, 2002.

———. "Navy Training Undercut by Fraudulent Reports" *DefenseWatch*, October 2, 2002.

———. "The Wrong Sub for New Warfare Era." *DefenseWatch*, October 4, 2004.

———. "Tomorrow's Submarine Fleet: The Non-nuclear option" *DefenseWatch*, February 2, 2002.

"World University Rankings." *Times Higher Education Supplement* (online), November 5, 2004.

Yack, Patrick. "Retired Officer's Critique Gets Heave-Ho from Navy." *Times-Union*, June 27, 1982.

Yamamoto, H. "Recent Trends of Drug Abuse in Japan." *Annals of the New York Academy of Sciences*, no. 1025 (2004).

Zampini, Diego. "North Vietnamese Aces: Mig-17 and Mig-21 Pilots, Phantom and Thud Killers," Acepilots.com.

BOOKS, REPORTS, SEMINARS, AND MONOGRAPHS

Abrashoff, D. Michael. *Get Your Ship Together*. New York: Portfolio, 2004.

———. *It's Your Ship*. New York: Warner Books, 2002.

Adams, Michael C. C. *The Best War Ever: America and World War II*. Baltimore: Johns Hopkins University Press, 1994.

Alcott, James A. *Printed Media in the Year 2001*. Kansas City, Mo.: Midwest Research Institute, 1978.

Allen, Thomas B. *War Games: The Secret World of the Creators, Players, and Policy Makers Rehearsing World War III Today*. New York: McGraw-Hill, 1987.

Aloni, Shlomo. *Israeli Mirage and Nesher Aces*. Oxford: Osprey, 2004.
Anderson, Terry H. *The Movement and the Sixties*. New York: Oxford University Press, 1995.
Anelauskas, Valdas. *Discovering America as It Is*. Atlanta: Clarity, 1999. (The cover illustration is objectionable, but nevertheless this is a very well researched critique on American culture, education, and politics.)
Appleman, Roy E. *Disaster in Korea: The Chinese Confront MacArthur*. College Station: Texas A&M University Press, 1989.
Appy, Christian G. *Patriots: The Vietnam War Remembered from All Sides*. New York: Penguin, 2003.
Arkin, William F., and Joshua Handler. *Naval Nuclear Accidents at Sea*. Neptune Papers III. Washington, D.C.: Greenpeace, 1989. (An extract from this book can be viewed at www.greenpeace.org/raw/content/international/press/reports/naval-nuclear-accidents.pdf.)
Atkinson, Gerald L. *From Trust to Terror: Radical Feminism Is Destroying the U.S. Navy*. Camp Springs, Md.: Atkinson Associates, 1997).
"Background on NATO Flying Training in Canada (NFTC)." Ottawa: Department of National Defence, November 1, 2003, available at www.airtraining.forces.gc.ca/airtraining/nftc_background_e.asp.
Beurling, George F., and Leslie Roberts. *Malta Spitfire: The Diary of a Fighter Pilot*. London: Greenhill Books, 2002.
Beyer, Rick. *The Greatest War Stories Never Told*. New York: Collins, 2005.
Biddle, Stephen. *Military Power: Explaining Victory and Defeat in Modern Battle*. Princeton, N.J.: Princeton University Press, 2004.
Boileau, John. *Fastest in the World: The Saga of Canada's Revolutionary Hydrofoils*. Halifax, N.S.: Formac, 2004.
Brehm, Jack, and Pete Nelson. *That Others May Live*. New York: Crown, 2000.
Brock, Jeffry. *The Dark Broad Seas*. Vol. 1, *With Many Voices*. Toronto: McClelland and Stewart, 1981.
———. *The Thunder and the Sunshine*. Vol. 2, *With Many Voices*. Toronto: McClelland and Stewart, 1983).
"Buy Six Diesel-Electric Submarines for Antisubmarine Warfare Training (050-26)." In *Budget Options 2001* (online). Washington, D. C.: Congressional Budget Office, February 2001.
Bruns, Roger. *Almost History*. New York: Hyperion, 2000.
Campagna, Palmiro. *Storms of Controversy: The Secret Avro Arrow Files Revealed*. Toronto: Stoddart, 1997.

Cerf, Christopher, and Victor Navasky. *The Experts Speak*. New York: Villard, 1998.
Chancellor, John. *Peril and Promise: A Commentary on America*. New York: HarperPerennial, 1991.
Citizens against Government Waste. *The Pig Book: How Government Wastes Your Money*. New York: Thomas Dunne Books, 2005.
Claire, Rodger. *Raid on the Sun*. New York: Broadway, 2004.
Clancy, Tom. *Carrier*. New York: Berkley Books, 1999. (Recommended only for middle school students doing term papers; adults with critical reasoning skills will quickly recognize it as a propaganda piece.)
———. *Fighter Wing*. New York: Berkley Books, 1995. (Entertaining, but not recommended because of its blatantly jingoistic tangents.)
———. *The Hunt for Red October*. New York: Berkley Books, 1999. (See the note above for his book *Fighter Wing*.)
———. *Submarine*. New York: Berkley Books, 1993. (See the above.)
Cockburn, Andrew. *The Threat: Inside the Soviet Military Machine*. New York: Vintage Books, 1984.
Cohen, Eliot A., and John Gooch. *Military Misfortunes: The Anatomy of Failure in War*. New York: Anchor, 2003.
Compton-Hall, Richard. *Submarine versus Submarine*. Toronto: Collins, 1988.
Cook, Theodore F. "Our Midway Disaster: Japan Springs a Trap, June 4, 1942." In *What If? The World's Foremost Military Historians Imagine What Might Have Been*, edited by Robert Cowley. New York: Berkley Books, 1999.
Cooper, Tom, and Farzad Bishop. *Iranian F-14 Tomcat Units in Combat*. Oxford: Osprey, 2004.
Coram, Robert. *Boyd: The Fighter Pilot Who Changed the Art of War*. New York: Back Bay Books, 2004.
Cote, Owen R. *The Third Battle: Innovation in the U.S. Navy's Cold War Struggle with Soviet Submarines*. Cambridge, Mass.: MIT Security Studies Program, March 2000.
Couch, Dick. *The Finishing School: Earning the Navy SEAL Trident*. New York: Crown, 2004.
Craven, John Pina. *The Silent War: The Cold War Battle beneath the Sea*. New York: Touchstone, 2001.
Deacon, Richard. *The Silent War: A History of Western Naval Intelligence*. London: Grafton Books, 1988.

Department of National Defence, *Report on Canadian Patrol Frigate Cost and Capability Comparison*, 7050-11-11 CRS. Ottawa: Chief Review Services, March 26, 1999.

Diehl, Alan E. *Silent Knights: Blowing the Whistle on Military Accidents and Their Cover-Ups.* Washington, D.C.: Brassey's, 2002.

DiMercurio, Michael, and Michael Benson. *The Complete Idiot's Guide to Submarines.* New York: Alpha. 2003. (Don't let the title bother you. This is a very highly detailed and well written book, superior to anything by Tom Clancy. The primary author is a former U.S. nuclear submarine officer.)

Dixon, Norman. *On the Psychology of Military Incompetence.* London: Pimlico, 1994.

Doenitz, Karl. *Memoirs: Ten Years and Twenty Days.* New York: Da Capo, 1997.

Donnelly, Elaine. "Legacy Project Launches Spin Campaign to Obscure Clinton Record on Military Readiness." January 24, 2002, available at cmrlink.org/terrorismwar.asp?docID=124.

Dorr, Robert F, and Warren Thompson. *Korean Air War.* St. Paul, Minn.: Motorbooks International, 2003.

Dorr, Robert F., et al. *Korean War Aces.* Oxford: Osprey, 1995.

Douthat, Ross Gregory. *Privilege: Harvard and the Education of the Ruling Class.* New York: Hyperion, 2005.

Dunnigan, James F., and Albert A. Nofi. *Dirty Little Secrets.* New York: Quill, 1990.

———. *Dirty Little Secrets of the Vietnam War.* New York: Thomas Dunne, 1999.

Eland, Ivan. *Subtract Unneeded Nuclear Submarines from the Fleet.* Foreign Policy Briefing 47. Washington, D.C.: Cato Institute, April 1, 1998.

Enright, Joseph F., with James W. Ryan. *Sea Assault.* New York: St. Martin's, 2000.

Fallows, James. *National Defense.* New York: Random House, 1981.

Fitzhenry, Robert I., ed. *The Harper Book of Quotations.* New York: Quill, 1993.

Friedman, George, and Meredith LeBard. *The Coming War with Japan.* New York: St. Martin's, 1991.

Freeman, Gregory A. *Sailors to the End: The Deadly Fire on the USS* Forrestal *and the Heroes Who Fought it.* New York: Perennial, 2002.

Fuchida, Mitsuo, and Masatake Okumiya. *Midway: The Battle That Doomed Japan, the Japanese Navy's Story.* Annapolis, Md.: Naval Institute Press, 2001.

Gabriel, Richard. *Military Incompetence: Why the American Military Doesn't Win.* New York: Noonday, 1985.

Gannon, Michael. *Operation Drumbeat: Germany's U-Boat Attacks along the American Coast in World War II.* New York: HarperPerennial, 1991.

German, Tony. *The Sea Is at Our Gates: The History of the Canadian Navy.* Toronto: McClelland and Stewart, 1990.

Gimblett, Richard. *Operation Apollo: The Golden Age of the Canadian Navy in the War against Terrorism.* Ottawa: Magic Light, 2004.

Goff, Stan. *Full Spectrum Disorder: The Military in the New American Century.* Brooklyn, N.Y.: Soft Skull, 2004.

Gordon, Yefim. *Mikoyan-Gurevich MiG-15.* Hinckley, U.K.: Aerofax, 2001, 65–73.

———. *Mikoyan MiG-31.* Hinckley, U.K.: Midland, 2005.

Gorshkov, S. G. *The Sea Power of the State.* Oxford: Pergamon, 1980.

Granatstein, J. L. *Who Killed the Canadian Military?* Toronto: Harper, 2004.

Gray, Chris Hables. *Peace, War, and Computers.* New York: Routledge, 2005.

Greider, William. *Fortress America.* New York: PublicAffairs, 1998.

Gross, Martin L. The *Conspiracy of Ignorance: The Failure of American Public Schools.* New York: Perennial, 2000.

Gutmann, Stephanie. *The Kinder, Gentler Military.* San Francisco: Encounter Books, 2000.

Hackworth, David H. *Hazardous Duty.* New York: Avon Books, 1996.

Hadley, Arthur T. *Straw Giant: America's Armed Forces: Triumphs and Failures.* New York: Avon, 1987.

Halberstadt, Hans. *U.S. Navy SEALS.* Osceola, Wis.: MBI, 1993.

Hallion, Richard *Storm over Iraq: Air Power and the Gulf War.* Washington, D.C.: Smithsonian, 1992.

Hammes, Thomas X. *The Sling and the Stone.* St. Paul, Minn.: Zenith, 2004.

Harris, Brayton. *The Navy Times Book of Submarines.* New York: Berkley Books, 1997.

Haydon, Peter. *The 1962 Cuban Missile Crisis: Canadian Involvement Reconsidered.* Toronto: Canadian Institute of Strategic Studies, 1993.

Hayes, Peter, Lyuba Zarsky, and Walden Bello. *American Lake: Nuclear Peril in the Pacific.* (Middlesex, UK: Penguin Books, 1987.

Henderson, William Darryl. *The Hollow Army: How the U.S. Army Is Oversold and Undermanned.* Westport, Conn.: Greenwood, 1990.

Hickam, Homer H., Jr. *Torpedo Junction.* New York: Dell, 1989.

Hirsh, Michael. *None Braver: U.S. Air Force Pararescuemen in the War on Terrorism.* New York: NAL Caliber, 2004.

Holland, Lauren. *Weapons under Fire*. New York: Garland, 1997).
Holland, W. J., ed. *The Navy*. Washington, D.C.: Hugh Lauter Levin, 2000.
Hoyt, Edwin. *War in the Pacific*. Vol. 1, *Triumph of Japan*. New York: Avon Books, 1990.
Huntington, Samuel. *The Soldier and the State*. Cambridge, Mass.: Belknap, 2003.
Huchthausen, Peter. *October Fury*. Hoboken, N.J.: John Wiley and Sons, 2002.
——, Igor Kurdin, and R. Alan White. *Hostile Waters*. New York: St. Martin's Paperbacks, 1997.
Hunter, Jamie. *Sea Harrier: The Last All-British Fighter*. Hinckley, U.K.: Midland, 2005.
Isenberg, David. *The Illusion of Power: Aircraft Carriers and U.S. Military Strategy*. Cato Policy Analysis 134. Washington, D.C.: Cato Institute, 1990.
Isenberg, Michael T. *Shield of the Republic: The United States Navy in an Era of Cold War and Violent Peace*. Vol. 1. New York: St. Martin's, 1993.
Johnson, Chalmers. *The Sorrows of Empire: Militarism, Secrecy, and the End of the Republic*. New York: Metropolitan Books, 2004.
Johnson, Donald. *It Wasn't Just a Job: It Was an Adventure*. Lincoln, Neb.: Writers Club, 2002.
Kaplan, Fred. "The Little Plane That Could Fly, If the Air Force Would Let It." In *More Bucks, Less Bang: How the Pentagon Buys Ineffective Weapons*, edited by Dina Rasor. Washington, D.C.: Fund for Constitutional Government Project on Military Procurement, 1983.
Karam, Andrew. *Rig Ship for Ultra-Quiet*. Sydney, Australia: Sid Harta, 2002.
Katz, Samuel M. *Israel's Air Force*. Osceola, Wis.: Motorbooks, 1991.
Keegan, John. *Six Armies in Normandy*. New York: Penguin, 1994.
Kernan, Alvin. *The Unknown Battle of Midway: The Destruction of the American Torpedo Squadrons, June 4, 1942*. New Haven, Conn.: Yale University Press, 2005.
Kick, Russ, ed. *Everything You Know Is Wrong*. New York: Disinformation, 2003.
Knox, J. H. W. "An Engineer's Outline of RCN History: Part II." In *RCN in Retrospect 1910–1968*, edited by James A. Boutilier. Vancouver: University of British Columbia Press, 1982.
Langton, Christopher. *The Military Balance 2003, 2004*. London: Oxford University Press, 2003.
Lansdown, John. *With the Carriers in Korea*. Manchester, UK: Crecy, 1997.

Lehman, John. *Command of the Seas*. New York: Charles Scribner's Sons, 1988.

Lehrman, Lewis E. *Address to Garrison Forest School*. Owings Mills, Md.: Lehrman Institute, September 23, 1993.

Levinson, Jeffrey L., and Randy L. Edwards. *Missile Inbound: The Attack on the Stark in the Persian Gulf*. Annapolis, Md.: Naval Institute Press, 1997.

Lindsey, Forrest R. "Nagumo's Luck." In *Rising Sun Victorious*, edited by Peter G.. Tsouras. London: Greenhill Books: 2001. (If nothing else, given the ongoing war raging in Europe and in the Atlantic, the very plausible albeit theoretical Japanese successes described in this book could easily have altered the course of U.S. naval history away from its current "big carrier" tangent.)

Lorber, Azriel. *Misguided Weapons: Technological Failure and Surprise on the Battlefield*. Washington, D.C.: Brassey's, 2002.

Lord, Walter. *Incredible Victory: The Battle of Midway*. Short Hills, N.J.: Burford Books, 1998.

Lovitz, Barbara E. "Making the Implicit Explicit: A Conceptual Approach for Assessing the Outcomes of Doctoral Education." Paper presented at the ASHE Symposium on Assessing Doctoral Students. Sacramento, California, November 2004.

Lund, W. G. D. "The Royal Canadian Navy's Quest for Autonomy in the North West Atlantic: 1941–1943." in *RCN in Retrospect 1910–1968*, edited by James A .Boutilier. Vancouver: University of British Columbia Press, 1982.

Luvaas, Jay. *Napoleon on the Art of War*. New York: Touchstone, 1999.

Macgregor, Douglas A. *Breaking the Phalanx*. Westport, Conn.: Praeger, 1997.

Marcinko, Richard, and John Weisman. *Rogue Warrior*. New York: Pocket Books, 1992.

Marolda, Edward J., and Robert J. Schneller Jr. *Shield and Sword: The United States Navy and the Persian Gulf War*. Washington, D.C.: Government Reprints Press, 2001.

McCaffery, Dan. *Air Aces: The Lives and Times of Twelve Canadian Fighter Pilots*. Toronto: Lorimer, 1990.

McManners, Hugh. *Top Guns*. London: Network Books, 1996.

Miller, David. *The Illustrated Directory of Submarines of the World*. London: Salamander, 2002.

Miller, Duncan, and Sharon Hobson. *The Persian Excursion: The Canadian Navy in the Gulf War*. Clementsport, N.S.: Canadian Peacekeeping, 1995.

Miller, Nathan. *War at Sea.* New York: Oxford University Press, 1995.
Milner, Marc. *North Atlantic Run: The Royal Canadian Navy and the Battle for the Convoys.* Toronto: University of Toronto Press, 1994.
Moore, Robert. *A Time to Die.* New York: Three Rivers, 2002.
Morin, Jean H., and Richard H. Gimblett. *Operation Friction: The Canadian Forces in the Persian Gulf.* Toronto: Dundurn, 1997.
Morison, Samuel Eliot. *The Battle of the Atlantic: September 1939–May 1943.* Champaign: University of Illinois Press, 2001.
———. *The Rising Sun in the Pacific 1931–April 1942.* Edison, N.J.: Castle Books, 2001.
Moskos, Charles. "Armed Forces in a Warless Society." In *War,* edited by Lawrence Freedman. Oxford: Oxford University Press, 1994.
Naylor, Sean. *Not a Good Day to Die: The Untold Story of Operation Anaconda.* New York: Berkley Books, 2005.
Neale, Jonathan. *A People's History of the Vietnam War.* New York: New Press, 2003.
Nichols, John B., and Barrett Tillman. *On Yankee Station.* Annapolis, Md.: Naval Institute Press, 2001.
Nordeen, Lon. *Fighters over Israel.* New York: Orion Books, 1990.
O'Connell, Robert L. *Sacred Vessels: The Cult of the Battleship and the Rise of the U.S. Navy.* New York: Oxford University Press, 1991.
Orita, Zenji, and Joseph D. Harrington. *I-Boat Captain.* Canoga Park, Calif.: Major Books, 1976.
Overy, Richard. *Why the Allies Won.* New York: W. W. Norton, 1995.
Padfield, Peter. *War beneath the Sea.* New York: John Wiley and Sons, 1998.
Parenti, Michael. *Superpatriotism.* San Francisco: City Light Books, 2004.
Peniston, Bradley. *Around the World with the U.S. Navy: A Reporter's Travels.* Annapolis, Md.: Naval Institute Press, 1999.
Peters, Charles. *How Washington Really Works.* Cambridge: Perseus, 1993.
Pillsbury, Michael. *China Debates the Future Security Environment* (online). Washington, D.C.: National Defense University Press, 2000.
Piper, Joan L. *A Chain of Events.* Dulles, Va.: Brassey's, 2001.
Polmar, Norman, and K. J. Moore. *Cold War Submarines.* Dulles, Va.: Brassey's, 2004.
Prange, Gordon. *At Dawn We Slept.* New York: Penguin, 1991.
Pugliese, David. *Canada's Secret Commandos.* Ottawa: Esprit de Corps Books, 2002.
Reeve, John, and David Stevens, eds. *The Face of Naval Battle.* Crows Nest, NSW: Allen and Unwin, 2003.

Regan, Geoffrey. *The Brassey's Book of Naval Blunders*. Washington, D.C.: Brassey's, 2000.

Rendall, Ian. *Rolling Thunder: Jet Combat from World War II to the Gulf War*. New York: Dell, 1997.

Riccioni, Everest. *Description of Our Failing Defense Acquisition System*. Washington, D.C.: Project on Government Oversight, March 8, 2005.

———. *Is the Air Force Spending Itself into Unilateral Disarmament?* Washington, D.C.: Project on Government Oversight, August 2001.

Richards, Chester. *A Swift, Elusive Sword*. Washington, D.C.: CDI, 2003.

Robb, David L. *Operation Hollywood: How the Pentagon Shapes and Censors the Movies*. Amherst, N.Y.: Prometheus Books, 2004.

Rose, Elihu. "The Case of the Missing Carriers." in *What If? The World's Foremost Military Historians Imagine What Might Have Been*, edited by Robert Cowley. New York: Berkley Books, 1999.

Russett, Bruce M. *No Clear and Present Danger: A Skeptical View of the United States Entry into World War II*. Boulder, Colo.: Westview, 1997.

Sadkovich, James J., ed. *Reevaluating Major Naval Combatants of World War II*. Westport, Conn.: Greenwood, 1990.

Scammell, Henry. *Giantkillers: The Team and the Law That Help Whistleblowers Recover America's Stolen Billions*. New York: Atlantic Monthly, 2004.

Schank, John F., et al. *Finding the Right Balance: Simulator and Live Training for Navy Units*. Santa Monica, Calif.: RAND, 2002.

Simpson, Howard R. *The Paratroopers of the French Foreign Legion*. Dulles, Va.: Brassey's, 1999.

Smith, Hedrick. *The Power Game*. New York: Ballantine Books, 1988.

Smith, Perry M. *Assignment Pentagon: How to Excel in Bureaucracy*. Dulles, Va.: Brassey's, 2002.

Snowie, Alan J. *The Bonnie:* HMCS Bonaventure. Richmond Hill, B.C.: Firefly Books, 1987. (This book provides detailed information on the successes of the *Bonaventure* and her ability to outperform American *Essex*-class ASW carriers.)

Sontag, Sherry, and Christopher Drew. *Blind Man's Bluff*. New York: HarperPaperbacks, 1998.

Soward, Stuart E. *Hands to Flying Stations*, Vol. 2. Victoria, B.C.: Neptune, 1995.

Spector, Ronald. *At War at Sea: Sailors and Naval Combat in the Twentieth Century*. New York: Penguin Books, 2001.

———. *Eagle against the Sun*. New York: Vintage, 1985.

Spencer, Jack. *The Facts about Military Readiness*. Backgrounder 1394. Washington, D.C.: Heritage Foundation, September 15, 1999.

Spinney, Franklin C. *Defense Facts of Life*. Boulder, Co.: Westview, 1985.

Stevenson, James *The Pentagon Paradox: The Development of the F-18 Hornet*. Annapolis, Md.: Naval Institute Press, 1993.

Stillion, John. *Blunting the Talons: The Impact of Peace Operations Deployments on USAF Fighter Crew Combat Skills*. RGSD-147. Santa Monica, Calif.: RAND, 1999.

Stubbing, Richard A., and Richard A. Mendel. *The Defense Game*. New York: Harper and Row, 1986.

Sullum, Jacob. *Saying Yes: In Defense of Drug Use*. New York: Tarcher/Putnam: 2003.

Sykora, Dale. "An Operational Perspective of Submarine Evasion Operations." Presentation at the 4th International New Horizons in Search Theory Workshop: Investigating Hider Theory, Newport R.I., April 2004, available at old.alidade.net/events/040427_SearchTheory/05-submarine-evasion.htm.

Tagaya, Osamu. *Imperial Japanese Naval Aviator 1937–45*. Wellingborough, U.K.: Osprey, 2003.

Thompson, Roger. *Brown Shoes, Black Shoes, and Felt Slippers: Parochialism and the Evolution of the Post-War U.S. Navy*. (Newport: U.S. Naval War College, 1995).

Thompson, William C. *A Glimpse of Hell: The Explosion on the USS Iowa and its Cover-Up*. New York: W. W. Norton, 1999.

Thorburn, Doug. *How to Spot Hidden Alcoholics: Using Behavioral Clues to Recognize Addiction in Its Early Stages*. Northridge, Calif.: Galt, 2004.

Timperlake, Edward, and William C. Triplett III. *Red Dragon Rising: Communist China's Military Threat to America*. Washington, D.C.: Regnery, 2000.

Toffler, Alvin, and Heidi Toffler. *War and Anti-War*. New York: Warner Books, 1993.

Tomlison, Thomas M. *The Threadbare Buzzard: A Marine Fighter Pilot in WWII*. St. Paul, Minn.: Zenith, 2004.

Toperczer, Istvan. *MiG-21 Units of the Vietnam War*. Oxford: Osprey, 2001.

Trask, David F. *The AEF & Coalition Warmaking, 1917–1918*. Lawrence: University Press of Kansas, 1993.

Truscott, Peter. *Kursk*. London: Pocket Books, 2003.

Tyler, Patrick. *Running Critical: The Silent War, Rickover, and General Dynamics*. New York: Harper and Row, 1986.

Vandergriff, Donald E. *Raising the Bar: Creating and Nurturing Adaptability with the Changing Face of War.* Washington, D.C.: Center for Defense Information, 2006.
van der Vat, Dan. *The Atlantic Campaign.* Edinburgh: Birlinn, 2001.
———. *The Pacific Campaign: The U.S.-Japanese Naval War 1941–1945.* New York: Touchstone, 1991.
Vistica, Gregory L. *Fall from Glory: The Men Who Sank the U.S. Navy.* New York: Simon and Schuster, 1997.
Waller, Douglas C. *Big Red: The Three Month Voyage of a Trident Nuclear Submarine.* New York: HarperTorch, 2001.
Walmer, Max. *An Illustrated Guide to Modern Elite Forces.* New York: Prentice Hall, 1986.
Ward, Sharkey. *Sea Harrier over the Falklands.* London: Cassell Military, 2006.
Weir, Gary E., and Walter J. Boyne. *Rising Tide: The Untold Story of the Russian Submarines That Fought the Cold War.* New York: Basic Books, 2003.
Wilcox, Robert K. *Black Aces High.* New York: St. Martin's Paperbacks, 2002.
———. *Scream of Eagles.* New York: Pocket Books, 2005.
———. *Wings of Fury.* New York: Pocket Books, 2004.
Williams, Andrew. *The Battle of the Atlantic.* New York: Basic Books, 2003.
Wilson, George C. *Supercarrier.* New York: Berkley Books, 1986.
Woolner, Derek. *Getting in Early: Lessons of the Collins Submarine Program for Improved Oversight of Defence Procurement.* Research Paper 3 2002-02. Canberra, ACT: Department of the Parliamentary Library, Foreign Affairs, Defence and Trade Group, 18 September 2000.
Z., Mickey. *The Seven Deadly Spins.* Monroe, Me.: Common Courage, 2004.
Zepezauer, Mark. *Take the Rich Off Welfare.* Cambridge, Mass.: South End, 2004.
Zumwalt, Elmo, and [Lt.] Elmo Zumwalt. *My Father, My Son.* New York: Dell, 1987.

GAO REPORTS

Defense Acquisitions: Comprehensive Strategy Needed to Improve Ship Cruise Missile Defense. Report GAO/NSIAD-00-0149. Washington, D.C.: GAO, July 2000.
Statement of Frank C. Conahan, Director National Security and International Affairs Division before the Subcommittee on Legislation and National Security Committee on Government Operations, House of Representatives, on

Readiness of Navy Tactical Air Forces. Washington, D.C.: GAO, November 2, 1983.
Weapons Testing: Quality of DOD Operational Testing and Reporting. Report GAO/PEMD-88-32BR. Washington, D.C.: July 1998.

ONLINE RESOURCES

216.239.63.104/search?q=cache:iBVfkb47JpAJ:www.nwc.navy.mil/usnhdb/losses_war.asp+sunk+by+a+submarine&hl=en. War losses: a list of U.S. naval ship losses in World War II, compiled by the U.S. Naval War College.

www.dutchsubmarines.com. An excellent site maintained by Dutch submariners. It has extensive entries on Dutch successes against U.S. Navy carriers and some startling periscope photos as well.

www.globalsecurity.org. A very authoritative site that features detailed information on the U.S. Navy, its ships, its bases, and its equipment. Maintained by John Pike.

www.scramble.nl/pk.htm. The Dutch Aviation Society publishes very detailed information on world air forces on this website. Very professional and informative. Details of the Pakistani Air Force's successes against the Indian and Israeli air forces are provided.

www.theaerodrome.com/contrib/vs.html. This website provides detailed information on all the ace fighter pilots of World War I.

www.f-16.net/interviews.html. See the interviews with Maj. Russell Prechtl and Maj. Gregory Stroud for more data on the F-16 and for comparisons of the U.S. Navy and Air Force flying programs.

www.atlantic.drdc-rddc.gc.ca/about/achievements_e.shtml. Information about the CANTASS system.

www.royal-navy.mod.uk/static/pages/450.html. Royal Navy Mine countermeasures.

www.globalissues.org/Geopolitics/ArmsTrade/Spending.asp#InContextU.S.MilitarySpendingVersusRestoftheWorld. U.S. military spending vs. that of other countries.

www.iwm.org.uk/upload/package/8/atlantic/can3942.htm. The Imperial War Museum's Web site on the RCN's contribution to the Battle of the Atlantic. *www.ussdevilfish.com/interv.htm.* Interview with Michael DiMercurio by Christy Tillery French.

www.navsource.org/archives/03/0301132.jpg. Image of North Vietnamese postage stamp commemorating the sinking of USS *Card*.

www.blueangels.navy.mil/staticFiles/team.html, and *www.blueangels.navy.mil/geninfo/fa_aline.html.* Blue Angels information.

www.brainyquote.com, and *en.thinkexist.com/.* The wit and humor of the ages.

www.chinfo.navy.mil/navpalib/ships/carriers/histories/cv03-saratoga/cv03-saratoga.html. USS *Saratoga* information.

www.dtic.mil/doctrine/jel/new_pubs/jp1_02.pdf. Department of Defense *Dictionary of Military and Associated Terms* amended May 2005.

www.snowbirds.forces.gc.ca/faq_e.asp#3. Snowbirds information.

www.ibiblio.org/hyperwar/U.S.N/U.S.NatWar/U.S.N-King-2.html. "U.S. Navy at War 1941–1945 (Official Reports by Fleet Admiral Ernest J King, U.S.N.)"

dictionary.cambridge.org. Cambridge dictionaries.

www.physics.northwestern.edu/classes/2001Fall/Phyx135-2/19/emp.htm. Electromagnetic-pulse weapons.

www.cv6.org/1941/1941.htm. USS *Enterprise* (CV 6).

www.globalsecurity.org/military/agency/navy/cargru5.htm. Carrier Group Five.

people.hofstra.edu/geotrans/eng/ch1en/conc1en/carprod1950-1999.html. "Automobile Production, United States, Japan and Germany, 1950–2004 (in millions)."

www.natotigers.org/ntmpage/ntm_06.html. "NATO Tiger Meet Winners."

home.att.net/~jbaugher1/f5_28.html. "Northrop F-5E Tiger II for the U.S. Navy," by Dr. Joe Baugher, 2003.

t2web.amnesty.r3h.net/library/Index/ENGASA170432004?open&of=ENG-CHN. "China: Fear of Imminent Execution: Ma Weihua."

en.wikipedia.org/wiki/Oberon_class_submarine. Oberon-class submarines.

www.drdc-rddc.gc.ca/about/history_e.asp. Variable-depth sonar and antigravity suits.

www.globalsecurity.org/military/world/spending.htm. Worldwide military expenditures.

en.wikipedia.org/wiki/F-14_Tomcat. Wikipedia article on the F-14 Tomcat.

en.wikipedia.org/wiki/Iran_Air_Flight_655. Wikipedia article on Iran Air 655 and the USS *Vincennes.*

www.f-16.net/f-16_users_article24.html. U.S. Navy F-16s.

www.tailhook.org/AVSLANG.htm. Aviator's slang.

SPEECHES

Bowman, Skip. "Honoring Our Own Greatest Generation." Chattanooga Armed Forces Day Luncheon. Chattanooga, Tennessee, May 6, 2005.

Cellucci, Paul. "Remarks by Ambassador Paul Cellucci to the American Assembly." February 4, 2005. Arden House, Harriman, New York, available at www.usembassycanada.gov/content/content.asp?section=emb-consul&subsection1=cellucci_speeches&document=cellucci_020405.

Polmar, Norman. Statement to the House Military Procurement Subcommittee. Washington, D.C., March 18, 1997.

RADIO INTERVIEWS

"Paul Beaver Discusses the U.S. Navy's Public Relations Program Compared with Others around the World." NPR *Weekend Edition: Sunday with Lisa Simeone*. February 17, 2001.

Interview with Thomas B. Allen. NPR *Talk of the Nation* with Neal Conan, January 8, 2003.

TELEPHONE INTERVIEW

Interview with Capt. Dean Knuth, USNR (Ret.), 10 March 2005.

E-MAIL CORRESPONDENCE

E-mails from Commander Peter Kavanagh, CF (Ret.), Lieutenant Colonel David Bashow, CF (Ret.), Capt. John L. Byron, USN (Ret.), Captain Jan Nordenman, Royal Swedish Navy (Ret.), Lieutenant Commander Aidan Talbott, RN, Dr. Andy Karam, Squadron Leader J. R. Sampson, RAAF (Ret.), Lt. Col. David Evans, USMC (Ret.), Jon Dougherty, Col. Everest Riccioni, USAF (Ret.), Lieutenant Colonel Pierre Rochefort, CF (Ret.), Major Lew Ferris, CF (Ret.), Major Leif Wadelius, CF (Ret.), Luka Novak, Roger Larsson, Craig Fosnock, and Lieutenant Commander Roland West, RCN (Ret.), in 2004, 2005, and 2006.

UNPUBLISHED PAPERS AND BOOK MANUSCRIPTS

Appleton, Daniel S. "Sailors in America's Naval Warships Are Not Prepared to Cope with the Violence of Battle."

Kelly, Christopher J. "The Submarine Force in Joint Operations." Research Report AU/ASCS/145-1998-04. Maxwell Air Force Base, Ala.: Air Command and Staff College, April 1998.

Knuth, Dean. "Lessons of Ocean Venture '81." 1981.

Riccioni, Everest. "A History and Military Analysis of the USAF F-22 Raptor Acquisition." Unpublished manuscript. 2005.

Shuger, Scott. *Navy Yes, Navy No.* Unpublished manuscript, 1988.
Walker, Samuel J. "Interoperability at the Speed of Sound: Canada-United States Aerospace Cooperation . . . Modernizing the CF-18 Hornet." Center for International Relations, Queen's University, Kingston, Ontario, Canada.

TV AND VHS REPORTS AND DOCUMENTARIES

Cran, William, and Ben Loeterman, producers. Frontline Documentary (PBS). "Return of the Great White Fleet." May 14, 1984.
Discovery Channel Documentary. "Building the Ultimate Submarine." 2003.
Discovery Channel Documentary. "Fleet Command." December 1997.
History Channel Documentary. "Heavy Metal: MiG-5 (Russian Stealth)." 2002.
Investigative Reports (with Bill Kurtis). "Out of the Gulf, into the New Navy." A&E Entertainment, 2000.
Investigative Reports (with Bill Kurtis). "Seven Minutes That Stunned the Navy." A&E Entertainment, 2000.
NOVA Documentary (PBS). "Submarine!" January 14, 1992.
NOVA Documentary (PBS). "Submarine: Steel Boats, Iron Men." 1989.
WABC-TV, New York. "Investigators Uncover Serious Flaws in Security at U.S. Navy Bases." October 24, 2000. Transcript available at abclocal.go.com/wabc/features/WABC_investigators_navybases.html.
Swan, Luke. "Great Planes: The F-104 Starfighter." AVI International (Australia), 1989.

Index

Abrashoff, Michael, 60, 106–7, 146, 149, 161
accidents, 143–45, 172–74
Adams, David, 4, 180
Aegis system, 176–77
Air Combat magazine, 125–26
air combat maneuvers, 108–37
    Canadian, 108–16
    Israeli, 116–20
Air National Guard, 137
Akula-class submarines, 42–43
Allen, Thomas B., 17, 63
Aloni, Shlomo, 119
American Civil Liberties Union, 166
Anderson, George W., 16, 59
Anelauskas, Valdas, 162–63
antisubmarine warfare (ASW), 40–62
Appleton, Daniel, 157
Aristophanes, 36
Arleigh Burke-class destroyers, 146, 148–49
Arnold J. Isbell, USS, 2
Arthur, Stanley, 38
Arthur W. Radford, USS, 153
ASDIC (Anti-Submarine Detection Investigation Committee), 52–53
ASW. *See* antisubmarine warfare (ASW)
asymmetric warfare, 95–96, 106–7
Atkinson, Gerald L., 163–64
Atwood, Margaret, 5
Aucoin, Joseph, 132
Augusta, USS, 92
Australian navy, 26–30
Axworthy, Thomas, 115

Baker, A. D., III, 45
Baker, Newton D., 65
Baker, Wilder D., 51
Base Realignment and Closure reports, 40
Bashow, David, 112–13, 130
Bay, Austen, 114–15
Bazely, HMS, 55–57
Beaver, Paul, 143
Bennington, Ashley, 33
Beurling, Buzz, 109–10
beyond visual range (BVR) targets, 126–27
Biddle, Stephen, 77–78
Big Red (Waller), 169
Bishop, Farzad, 86
Bismarck, Otto Von, 48
Blais, Jean Jacques, 22
Bloom, Howard, 82
Blue Angels, 150–52
Bonaventure, HMCS, 110–11, 171
Boomer, Walter, 38
Boorda, J. M., 43
Boyd, John, 129–30
Boyne, Walter J., 91, 112
Brahamaputra, INS, 171–72
Breaking the Phalanx (Macgregor), 106
Brittin, Burdick, 76
Brock, Jeffry, 48, 55–56, 75, 115, 140–41, 142
Brown, Harold, 102
Bulletin of the Atomic Scientists, 26
Burke, Arleigh, 57–58, 77
Burke, Edmund, 180
Burns, Jerry, 63, 117, 135, 165–66

Bush, George H. W., 12, 174
Bush, George W., 136
Byrne, Neil, 155
Byron, John L., 34–35, 102

Canada. *See* Royal Canadian Navy
Card, USS, 64–65
carriers, 81–93
    "invincibility" of, 64–69
    nuclear submarines vs., 57–58
    realistic appraisals of, 101–7
Carroll, Eugene, 103
Carter, Jimmy, 102
censorship, 63–65, 133
CF-104Gs, 111–12
CF-18s, 112
Chancellor, John, 146–47
Chang Mengxiong, 96
Chao, Luu Huy, 122
Chao-peng, Li, 34
Charles De Gaulle, French carrier, 134–35
Charlotte, USS, 62
Chilean navy, 31, 32, 125–29
China Debates the Future Security Environment (Pillsbury), 94
Chinese military, 82–83, 94–99
Churchill, Winston, 50
civilian ride-along program, 143–44
Claire, Rodger, 124
Clancy, Tom, 10–11, 34, 43, 72, 83, 113, 137, 168–75
Cloutier, Todd, 23
Coc, Nguyen Van, 124
Cochran, Donnie, 152
Cochrane, Wes, 151
Cockburn, Andrew, 33, 120–21, 125
Cohen, Eliot, 47, 48, 49
Cole, USS, 97–98, 99
Collins, Al, 107
Coming War with Japan, The (Friedman, Lebard), 145
Commander Task Force 24 (CTF 24), 54
communications equipment, 55–57, 145–46, 148
Compton-Hall, Richard, 14, 16, 22, 27, 34, 61–62, 171
Constellation, USS, 92–93, 152–53
Contractual Engineering Technical Service (CETS), 161–62

Cook, Theodore F., 69
Cooper, Tom, 86
Coral Sea, USS, 158
Coram, Robert, 9, 130
Corpus, Victor N., 78–79, 95
Cote, Owen, 42
Coyle, Frank G., 38
Craven, John, 90
Crawley, James W., 45
Crock, Stan, 132
Cuban Missile Crisis, 58–60
Cullen, Shay, 172
Cunningham, Randy "Duke," 124

Daly, Jack, 97
Davies, D. A., 113
defense budgets, 108–16, 136
Defense Officer Military Personnel Act (1980), 141
defensive system testing, 84–86
Delgado, James, 16
DeStasio, Robert, 130
Diehl, James G., 110, 134
diesel submarines, 10–40
    ASW measures and, 40–62
    lack of U.S., 36
    noise level of, 33–34, 41–44
    nuclear submarines vs., 24–40
    tracking/stalking by Soviet, 89–93
DiMercurio, Michael, 33, 34, 43, 172–73
Disaster in Korea: The Chinese Confront MacArthur (Appleman), 94
dishonesty, 13, 176–79
Dixon, Norman, 77
DOD Dictionary of Military and Associated Terms, 8
Doenitz, Karl, 46–47, 51
Dolphin, USS, 173
Donald, Kirkland H., 30
Dorsey, Jack, 114, 153
Dougherty, Jon, 167
Draz, Dave, 125
Driscoll, Willie, 124
Dunn, Robert F., 4
Dunnigan, James F., 64, 139
Dwight D. Eisenhower, USS, 2, 19–20
Dzhones, Pavel, 77

INDEX

education, 161–64
Edwards, Gordy, 111
Edwards, R. S., 47–48
Eisenhower, Dwight D., 15, 49, 180
Elkins, Al, 9–10
engineering, seamanship vs., 169–70
Enright, Joseph, 66
Enterprise, USS, 121
Ethell, Jeffrey, 125–26
Etzhold, Thomas, 63
Evans, David, 144
exercises, 8–14
    asymmetric warfare, 106–7
    diesel vs. nuclear submarines in, 24–40
    evaluation data from, 13
    force-on-force, 9–10
    free play, 9–10
    noise augmenters in, 40–44
    Northern Star, 22–23
    Ocean Venture, 17–21
    Operation Mainbrace, 15–17
    ships destroyed in, 185–86
    Uptide, 17
Exocet missiles, 85

F/A-18E/F Super Hornets, 132
Falkland Islands war (1982), 14
Fallows, James, 14, 85–86, 128
Flanagan, William, 4
Fleet ASW Command, 45
Forrestal, USS, 19–20, 159
Franklin D. Roosevelt, USS, 15–16
Freeman, Gregory, 159
French navy, 134–35, 153–54, 155–56
Friedman, George, 145
friendly fire incidents, 20, 126, 174
F-14s, 129–32, 134–35
F-15s, 111–12
F-18s, 134–35, 153–54
Fuchida, Mitsuo, 70

Gabriel, Richard, 155, 161
Gannon, Michael, 50
George Washington, USS, 143
German, Tony, 55, 58, 139–40
Gertz, Bill, 92–93
Gibson, Robert L. "Hoot," 122
Gillcrist, Paul, 132
Gimblett, Richard, 32, 145, 149–50

GlobalSecurity.org, 98
Goebbels, Josef, 54
Goff, Stan, 13, 106
Goldstein, Lyle, 44, 94–95, 105
Gooch, John, 47, 48, 49
Gordon, Yefim, 130
Gorshkov, S. G., 72–74, 80, 81, 159
Gotland-class submarines, 30, 45
Gough, Frank, 150–51
Granatstein, J. L., 75
"Great Planes: The Lockheed F-104 Starfighter," 111–12
Greeneville, USS, 143
Greenling, USS, 172
Greider, William, 78
Guadalcanal, 73–74
Guardfish, USS, 172
Gulf War
    Canadian pilots in, 112
Gunn, Lee F., 117
Gutmann, Stephanie, 64

Hackworth, David, 13
Haddock, USS, 173
Halifax-class frigates, 147–49
Hallion, Richard P., 120
Hammes, Thomas, 11, 77
Hardegen, Reinhard, 53
Harris, Brayton, 74
Hart, Gary, 21
Hartford, USS, 24–26
Hartmann, Eric, 128
Hawaii, Soviet submarine approach to, 90
Haydon, Peter, 58, 59
Hayes, Peter, 103
Heinl, Robert D., Jr., 165
Henderson, Ronald H., 153
Hickam, Homer, 49, 50–51
Hill, Robert, 28
Holland, Lauren, 177–78
Holland, W. J., 33, 83
Holloway, James, 84
Holzer, Robert, 32
Horne, Max, 53
Houlgate, Brian, 114
Hoyt, Edwin, 79–80
Huchthausen, Peter, 56, 59, 92
Hudson, Jason, 151
Hunter, Tim, 113

Huntington, Samuel, 178
Hussein, Saddam, 87, 105
Huxley, Aldous, 14

Independence, USS, 126, 153
Indo-Pakistani War, 123
intellectual dishonesty, 13
intelligence gathering/dissemination, 118–19, 152–53
International Air Power Review, 131
invincibility, 3–5, 19
Iowa, USS, 67–68, 174
Iran military, 86
Iraqi military, 85, 126
Isenberg, David, 86
Isenberg, Michael, 4, 19, 56, 138
Israeli military, 112, 116–20, 124–25, 171–72
It Wasn't Just a Job: It Was an Adventure (Johnson), 166
It's Your Ship (Abrashoff), 149

Japanese military, 7
 ASW in, 47, 49
 Battle of Midway and, 69–73
 diesel submarines of, 30–31
 semiconductors and, 146–47
 technology of, 145
Jerusalem Post, 117
jingoism, 3–5
John F Kennedy, USS, 22, 66–67, 84, 118, 140, 153
Johnson, Donald, 166
Johnson, Jay, 43
joint duty assignments, 155
Joyce, Maurice, 153
Judicial Watch, 97

Karam, Andy, 24, 30–31, 36, 41, 60, 99, 139, 163, 169, 170
Kavanagh, Peter, 24–26
Kaylor, Robert, 91–92
Keegan, John, 109
Kelton, Maryanne, 28
Kernan, Alvin, 71–72
King, Ernest J., 46, 48, 49, 51–52, 54
Kinman, Brent, 169–70
Kitty Hawk, USS, 11, 21, 66–67, 81–83, 84–85, 87–88
Knuth, Dean, 13, 17–21, 22

Kongo, Japanese, 12
Korean War, 133–34, 142
Kornukov, Anatoly M., 84–85
Kosovo, 112–13, 135
Krasnaya zvezda, 41–42
Kress, Bob, 132
Kurdin, Igor, 92
Kursk, Russian submarine, 91

La Jolla, USS, 173
Lake, Jon, 131
LaSilva, Ron, 31
leadership, 149, 161
LeBard, Meredith, 145
Lehman, John, 13, 18, 46, 65, 102, 104, 118, 134, 160, 161
"Lessons of Ocean Venture '81" (Knuth), 19–20
Libyan air force, 86–87
Lindsey, Forrest R., 74, 75
Littoral Combat Ships, 39–40
Lockwood, Charles A., Jr., 23
Lord, Walter, 71
Los Angeles-class ships, 170
Lott, Trent, 177
Lovitz, Barbara, 163
Lund, W. D. G., 53–54

Macgregor, Douglas, 106, 181–84
Mahan, Alfred Thayer, 178
MALABAREX (2003), 171–72
March, Dan, 37
Marcinko, Richard, 28, 98–99
Marolda, Edward, 37, 38, 103
Marshall, George C., 51, 183
McCaffrey, Dan, 110
McCrone, James, 137
McGwire, Mike, 60
McMillan, Joel, 139
McVadon, Eric, 83
medals, competition for, 142–43
Mendenhall, Corwin, 24
Metzger, James, 28
Michigan, USS, 92
Midway, Battle of (1942), 69–74, 79
Midway, USS, 15–16
Mies, Richard W., 44
MiG-17s, 122
MiG-31s, 130–31
militarism, patriotism and, 1–2

Miller, Nathan, 52
Miller, Peter, 28
mine countermeasures, 36–40
Mitchell, Billy, 65
Mohr, Kaleun Johann, 50–51
Moore, Karen, 67
Moore, Mike, 16
Moorer, Thomas, 104–5
Moosally, Fred, 67
morale, 121, 158–67
Morin, Jean, 32
Morison, Samuel, 52, 55, 69–70
Moskos, Charles, 108
Mullen, Michael G., 181–83
multitasking, 149–57
Murley, Steven P., 156
Murray, William, 44, 94–95

National Defense (Fallows), 14
NAUs. *See* noise augmenters (NAUs)
Nautilus, USS, 159
Naval Institute Guide to Combat Fleets of the World, 45
Navy, The, 4
Navy Yes, Navy No (Shuger), 88
Naylor, Sean, 100
Nebraska, USS, 169
New York City, USS, 170
Nichols, John B., 118, 121
Nimitz, USS, 86–87, 165
Noble, Percy, 55
Noel, Martin A., Jr., 109
Nofi, Albert A., 64, 139–40
noise augmenters (NAUs), 40–44
NORAD (North American Air Defense), 111
Nordeen, Lon, 119, 128–29
Norris, Bob, 136–37
North Carolina, USS, 11
North Korea, USS *Pueblo* and, 68–69
Northern Star exercise (1989), 22–23
Not a Good Day to Die: The Untold Story of Operation Anaconda (Naylor), 100
nuclear accidents, 172–75

O'Brien, J. C., 58
Ocean Venture (1981), 17–21
O'Connell, Robert L., 70
October Fury (Huchthausen), 59
October War (1973), 105

officers, 13, 140, 141–43, 149, 158–61
O'Hanlon, Michael, 94
Ohio, USS, 97
Okanagan, HMCS, 21–22
Olympia, USS, 28
On Watch (Zumwalt), 183
O'Neal, Kevin M., 152
Onondaga, HMCS, 24
Operation Desert Storm, 31–32, 37–38, 87, 126, 139
Operation Enduring Freedom, 39, 150
Operation Hollywood: How the Pentagon Shapes and Censors the Movies (Robb), 133
Operation Mainbrace (1952), 15–17
Operation Provide Comfort (1994), 126
Oriskany, USS, 121
Orita, Zenji, 47, 70, 73–74, 75
over-confidence, 178
overmanning, 149–57
Overy, Richard, 74

P-3 Orions, 44–45
Padfield, Peter, 46
Pakistani navy, 31
Parche, USS, 165
Parenti, Michael, 2
Parkin, Russell, 48–49
Pasadena, USS, 171–72
patriotism, militarism and, 1–2, 3–5
Patton, George S., 54
Pearl Harbor attack, 79–80, 88
Pedersen, Dan, 121
Pegasus, USS, 107
Peniston, Bradley, 9
Perry, Raymond, 140, 141, 155
Peterson, Gordon, 156–57
Petryk-Bloom, Dianne Star, 82
Phantoms, 122
Pharris, USS, 174
Phoenix missiles, 85–86
physical fitness, 167
Pillsbury, Michael, 94
Plunger, USS, 99, 169
Polmar, Norman, 22, 102
pork-barrel politics, 177–79
Powell, Colin, 113

Powers, Rod, 162
Powis, Jonathan, 61–62
Prange, Gordon, 13
predominance, as measure of success, 77–79
Princeton, USS, 37
promotion system, 13, 140–43
public affairs officers (PAOs), 64
Pueblo, USS, 68–69

racism, 158–61
Rafales, 134–35
Raising the Bar: Creating and Nurturing Adaptability with the Changing Face of War (Vandergriff), 141
RAND Corporation, 153–54, 155–56
Ranger, USS, 159
readiness, 66–67, 139–40, 155–56
Reagan, Ronald, 67, 103
Regan, Geoffrey, 54, 67–68
Rendell, Ivan, 120
Riccioni, Everest, 85, 108, 110, 127
Richards, Chester, 85
Rickenbacker, Eddie, 109
Rickover, Hyman, 102, 155, 170, 171, 173
Rig Ship for Ultra Quiet (Karam), 169
Robb, David L., 133
Robinson, David, 58
Rochefort, Pierre, 128
Rogers, William C., 144
Roosevelt, Franklin D., 50, 51–52, 114, 174
Rose, Elihu, 80
Rowe, Steve, 135–36
Royal Air Force, 153–54
Royal Australian Air Force (RAAF), 88–89
Royal Canadian Navy
    ASW in, 46–57, 58–60
    aviation in, 109–16
    communications systems, 145–46
    Cuban Missile Crisis and, 58–60
    defense budget for, 108–16
    diesel vs. nuclear submarines and, 24–26
    escape training, 138–39
    exercise victories of, 20–22
    mine countermeasures by, 37
    nuclear submarine detection by, 92
    Patrol Frigates, 147–49
    Snowbirds, 150–52
    technology vs. training in, 122–23
    training in, 149–50
    in U.S. carrier strike groups, 31
Royal Navy (British)
    ASW in, 46–57
    in Desert Storm, 37–38
    escape training, 139
    mine countermeasures in, 37–40
    multitasking in, 149
    readiness, 139–40
    training, 156–57
Royal Netherlands Navy, 22–23
Rumsfeld, Donald, 143
Running Critical (Tyler), 170
Russett, Bruce, 14

sabotage, 158–59
Sabre Mark VIs, 122–23
Sadkovich, James J., 55
safety, 143–45, 172–75
Saito, Toshitsugu, 143
Sampson, J. R., 88–89
Samuel B. Roberts, USS, 143
San Francisco, USS, 155
San Francisco Chronicle, 116
Saratoga, USS, 11, 89, 121, 160
Sargo, USS, 173
Saxton, Jim, 98
S-3B Vikings, 44–45
Scammell, Henry, 176–77
Schlesinger, James, 101
Schneller, Robert, Jr., 37, 38, 103
Scorpion, USS, 172
Scott, Bob, 136
Sea Harriers, 131
Sea Power magazine, 156
SEALS, 99–100
Seaquist, Larry, 144–45, 155
Seawolf-class submarines, 43–44
security measures, 97–100
semiconductors, 146–47
Sepp, Kalev, 116
Sewell, Kenneth, 90
sexual harassment, 161
Sharpshooter magazine, 113
Sheehan, HMAS, 28
Shein, Keith, 114
Shield of the Republic (Isenberg), 138

INDEX

Shinano, Japanese supercarrier, 12, 66, 67
Shkval torpedo, 145
Shuger, Scott, 35, 36, 66, 76, 86–87, 88, 90–91, 102, 118–19, 123, 144, 152–53, 156, 168–69, 177
Simeone, Lisa, 143
Skipjack, USS, 57–58
Smith, Allan, 36–37
Smith, Charles, 145
Smith, Denny, 177
Snowbirds, 150–52
Soldier and the State, The (Huntington), 178
sonar, 52–53
SOSUS, 45, 92
Soviet/Russian military
    Kitty Hawke and, 81–83, 84–85
    missiles, 87–88, 145
    nuclear submarine detection, 92–93
    realistic comparisons with, 104–6
    safety, 172
    SS-N-22 Sunburn missiles, 82–83
    stalking by submarines, 89–93
    submarine noise level in, 40–44
    substance abuse in, 7
    training, 154
Soward, Stuart, 57–58, 110, 154
Spadefish, USS, 159
specialization, 149–57
Spector, Ronald, 55, 120, 154, 159, 164, 170
Spencer, Jack, 161
Spinney, Franklin, 126–27, 153–54, 156
Sprey, Pierre, 128
SS-N-22 Sunburn missiles, 82–83
Stark, USS, 85, 143
Sterling, Jack, 114
Stevenson, James P., 122
Stimson, Henry, 49
Stroud, Gregory, 137
submarine, 10–40
submarine escape training, 138–39
substance abuse, 7, 164–67
Sullum, Jacob, 166
Sun Tzu, 94
Sun-Sin, Yi, 77
Swedish navy, 30, 45
Sweetman, Bill, 132

Taciturn, HMS, 16
Tae-Hee, Yoon, 163
Taiwanese navy, 34, 94–95
Talbott, Aidan, 146, 149
Tandem Thrust (1999), 9–10
Tate, Rob, 109–10
Teague, T. S., 83
technology, 12–13, 145–49
    air combat and, 108–18
    F-14s, 129–32
    training vs., 121–24
terrorism, 97–99
Tet offensive, 88
Theodore Roosevelt, USS, 117, 134–35
Thompson, William II, 67–68
Thorburn, Doug, 166
Threat, The: Inside the Soviet Military Machine (Cockburn), 33
Thresher, USS, 172
Ticonderoga, USS, 17
Tiger Meet of the Americas (TMOTA), 113
Tillman, Barrett, 118, 121
Times Higher Education Supplement, 163
Tomlinson, Thomas, 111
Top Gun course, 123–24, 132–35, 137
Top Gun (movie), 132–35
Toyka, Viktor, 34
training, 138–57
    education and, 161–64
    Netherlands, 23
    pilot, 113–14
    specialization in, 149–57
    technology vs., 121–24
    Top Gun, 123–24, 132–35
    US Air Force vs. Navy, 136–37
Truscott, Peter, 91, 139
Turner, Stansfield, 102–3
Twain, Mark, 79
Tyler, Patrick, 81, 89–90, 170, 173

underestimating enemies, 10–11
UNITAS exercise (1998), 8–9
Unknown Battle of Midway, The: The Destruction of the American Torpedo Squadrons (Kernan), 71–72
Uptide (1973), 17
U.S. Air Force, 2
    Canadian pilots compared with, 108–16

vs. Navy training, 136–37
U.S. Naval Academy, 163–64
U.S. Navy
   in air combat maneuvers, 108–37
   ASW in, 46–57
   censorship by, 63–65
   flight time in, 111–12
   intelligence gathering/
      dissemination by, 118–19
   lax security in, 97–100
   Lebanon raid, 118
   materiel condition of, 67–68
   overestimations by, 119
   political deception by, 176–79
   poor design of ships in, 55–57
   promotion system, 13, 140–43
   readiness of, 66–67
   reforms for, 179–80
   reliance of on allies, 36–40
   SEALS, 99–100
   submarines in WWII, 74–75
   support ships, 103–4
   testing practices of, 84–86
   Top Gun course, 123–24, 132–35
   turnover rates in, 24
   vs. Air Force training, 136–37
   waste in, 116
U.S. News & World Report, 32, 33

Van der Vat, Dan, 49, 80
Van Riper, Paul, 105–6
VandeLinde, David, 163
Vandergriff, Donald E., 100, 141
Victor III, Russian submarine, 42–43
Victoria-class submarines, 61–62
Vietnam War, 6, 158–61
   censorship in, 64–65
   competition for medals in,
      142–43
   Tet offensive, 88
   training in, 121–24
Vincennes, USS, 127, 133, 143, 144–45
Vistica, Greg, 121, 165

WABC-TV investigation, 97–98
Wajsman, Beryl P., 148
Walker, Samuel, 113

Waller, Douglas C., 169
Waller, HMAS, 28
War and Anti-War (Seaquist), 144–45
War Games (Allen), 63
Ward, N. D. "Sharkey," 131
Washington Times, 11, 88, 117
Wasp, USS, 15–16
waste, 116
Watkins, James, 42, 98
"We Are Not Invincible" (Adams), 4
weapons testing, 84–86
Weir, Gary E., 91
West, Roland, 110
Wieslander, Gumar, 30
Wilcox, Robert, 86, 129, 132
Wilde, Oscar, 134
William D. Porter, USS, 174
Williscroft, Robert, 34, 153
Wilson, George C., 118
Wilson, Woodrow, 52
Wings of Fury (Wilcox), 129
Winnipeg, HMCS, 150
Winnipeg Free Press, 21–22
Winston S. Churchill, USS, 156
Woolner, Derek, 28
World War I, Canadian pilots in, 109
World War II, 6
   ASW in, 46–57
   Battle of the Atlantic, 53–55
   Canadian pilots in, 109–10
   carriers sunk by diesel submarines in,
      11–12
   convoys in, 51–52
   Guadalcanal, 73–74
   Midway, 69–74
   Pearl Harbor, 79–80, 88
   uncoded messages in, 59
WorldNetDaily, 81–82

Yeager, Chuck, 118
Yom Kippur War, 124–25

Zanti, Guy W., 156–57
Zepezauer, Mark, 116
Zhukov, Yuri M., 105
Zumwalt, Elmo, 103, 104, 105, 142–43,
      160, 170, 183

About the Author

ROGER THOMPSON is a fellow of the Inter-University Seminar on Armed Forces and Society, a member of the Research Committee on Armed Forces and Conflict Resolution of the International Sociological Association, and an internationally recognized authority on combat motivation, military sociology, total-force issues, and military bureaucratic politics. His seminal work on combat motivation in naval forces was endorsed by the U.S. Chief of Naval Operations, SACLANT, and CINCPACFLT; by the late Capt. Edward L. Beach, USN (Ret.), a best-selling novelist and submariner; and by the German, Australian, Chilean, Italian, and Spanish admiralties. His work in this area was also translated into Spanish under the authority of the commander in chief of the Armada de Chile, Almirante Jorge Martinez Busch. In addition, Thompson received an Admiral's Medallion from the chief of staff of the Italian navy, Admiral Guido Venturoni, for his contribution to military sociology in 1993, and that same year Gen. Colin Powell also acknowledged his work in writing. His 1994 MA thesis *Brown Shoes, Black Shoes and Felt Slippers: Parochialism and the Evolution of the Post-War U.S. Navy* was published as a Strategic Research Department research report by the U.S. Naval War College in 1995 and again by the Mine Warfare Association in 1997.

Thompson has also published numerous military affairs essays in periodicals such as *Canada's Navy Annual, Air Force, Conference of Defence Associations Institute Forum, Canadian Defence Review, International Insights, Esprit de Corps,* and the *Defence Associations National Network News.* He currently lectures at Kyung Hee University in Korea.

THE NAVAL INSTITUTE PRESS is the book-publishing arm of the U.S. Naval Institute, a private, nonprofit, membership society for sea service professionals and others who share an interest in naval and maritime affairs. Established in 1873 at the U.S. Naval Academy in Annapolis, Maryland, where its offices remain today, the Naval Institute has members worldwide.

Members of the Naval Institute support the education programs of the society and receive the influential monthly magazine Proceedings and discounts on fine nautical prints and on ship and aircraft photos. They also have access to the transcripts of the Institute's Oral History Program and get discounted admission to any of the Institute-sponsored seminars offered around the country. Discounts are also available to the colorful bimonthly magazine *Naval History*.

The Naval Institute's book-publishing program, begun in 1898 with basic guides to naval practices, has broadened its scope to include books of more general interest. Now the Naval Institute Press publishes about seventy titles each year, ranging from how-to books on boating and navigation to battle histories, biographies, ship and aircraft guides, and novels. Institute members receive significant discounts on the Press's more than eight hundred books in print. Full-time students are eligible for special half-price membership rates. Life memberships are also available.

For a free catalog describing Naval Institute Press books currently available, and for further information about joining the U.S. Naval Institute, please write to:

Member Services
U.S. Naval Institute
291 Wood Road
Annapolis, MD 21402-5034
Telephone: (800) 233-8764
Fax: (410) 571-1703
Web address: www.navalinstitute.org